Lecture Notes
in Business Information Processing   553

Series Editors

Wil van der Aalst ⓘ, *RWTH Aachen University, Aachen, Germany*
Sudha Ram ⓘ, *University of Arizona, Tucson, USA*
Michael Rosemann ⓘ, *Queensland University of Technology, Brisbane, Australia*
Clemens Szyperski, *Microsoft Research, Redmond, USA*
Giancarlo Guizzardi ⓘ, *University of Twente, Enschede, The Netherlands*

LNBIP reports state-of-the-art results in areas related to business information systems and industrial application software development – timely, at a high level, and in both printed and electronic form.

The type of material published includes

- Proceedings (published in time for the respective event)
- Postproceedings (consisting of thoroughly revised and/or extended final papers)
- Other edited monographs (such as, for example, project reports or invited volumes)
- Tutorials (coherently integrated collections of lectures given at advanced courses, seminars, schools, etc.)
- Award-winning or exceptional theses

LNBIP is abstracted/indexed in DBLP, EI and Scopus. LNBIP volumes are also submitted for the inclusion in ISI Proceedings.

José María Moreno-Jiménez ·
Danielle Costa Morais · María Teresa Escobar ·
Alberto Turón
Editors

Human and Artificial Intelligence in Group Decision and Negotiation

25th International Conference on
Group Decision and Negotiation, GDN 2025
Zaragoza, Spain, June 15–18, 2025
Proceedings

Editors
José María Moreno-Jiménez
Universidad de Zaragoza
Zaragoza, Spain

Danielle Costa Morais
Universidade Federal de Pernambuco – UFPE
Recife, Brazil

María Teresa Escobar
Universidad de Zaragoza
Zaragoza, Spain

Alberto Turón
Universidad de Zaragoza
Zaragoza, Spain

ISSN 1865-1348 ISSN 1865-1356 (electronic)
Lecture Notes in Business Information Processing
ISBN 978-3-031-95220-3 ISBN 978-3-031-95221-0 (eBook)
https://doi.org/10.1007/978-3-031-95221-0

© The Editor(s) (if applicable) and The Author(s), under exclusive license
to Springer Nature Switzerland AG 2025

This work is subject to copyright. All rights are solely and exclusively licensed by the Publisher, whether the whole or part of the material is concerned, specifically the rights of translation, reprinting, reuse of illustrations, recitation, broadcasting, reproduction on microfilms or in any other physical way, and transmission or information storage and retrieval, electronic adaptation, computer software, or by similar or dissimilar methodology now known or hereafter developed.
The use of general descriptive names, registered names, trademarks, service marks, etc. in this publication does not imply, even in the absence of a specific statement, that such names are exempt from the relevant protective laws and regulations and therefore free for general use.
The publisher, the authors and the editors are safe to assume that the advice and information in this book are believed to be true and accurate at the date of publication. Neither the publisher nor the authors or the editors give a warranty, expressed or implied, with respect to the material contained herein or for any errors or omissions that may have been made. The publisher remains neutral with regard to jurisdictional claims in published maps and institutional affiliations.

This Springer imprint is published by the registered company Springer Nature Switzerland AG
The registered company address is: Gewerbestrasse 11, 6330 Cham, Switzerland

If disposing of this product, please recycle the paper.

Preface

The series of Annual International Conferences on Group Decision and Negotiation (GDN) have long served as a dynamic forum for the dissemination of cutting-edge research in the theory and practice of group decision-making and negotiation. These conferences provide an engaging environment where participants actively exchange, discuss, and critically evaluate emerging ideas in the field. Since its inception in 2000, the GDN conference series has been held annually, with the exception of 2011 and 2020. The chronological list of the conference locations is: 2000 (Glasgow, UK), 2001 (La Rochelle, France), 2002 (Perth, Australia), 2003 (Istanbul, Turkey), 2004 (Banff, Canada), 2005 (Vienna, Austria), 2006 (Karlsruhe, Germany), 2007 (Mont Tremblant, Canada), 2008 (Coimbra, Portugal), 2009 (Toronto, Canada), 2010 (Delft, The Netherlands), 2011 (cancelled), 2012 (Recife, Brazil), 2013 (Stockholm, Sweden), 2014 (Toulouse, France), 2015 (Warsaw, Poland), 2016 (Bellingham, USA), 2017 (Stuttgart, Germany), 2018 (Nanjing, China), 2019 (Loughborough, UK), 2020 (Toronto, Canada – conference cancelled due to the COVID-19 pandemic, though proceedings were published), 2021 (virtual, hosted from Toronto, Canada), 2022 (virtual), 2023 (Tokyo, Japan), and 2024 (Porto, Portugal).

The 25th International Conference on Group Decision and Negotiation (GDN 2025) was held from June 15 to 18, 2025, at the School of Business and Economics, University of Zaragoza, Spain. The conference attracted 99 submissions across nine main thematic streams within the GDN domain. Following a rigorous peer-review process in which each submission received three single-blind reviews, twelve papers were selected for inclusion in this volume, titled **Human and Artificial Intelligence in Group Decision and Negotiation.** These twelve papers are organized into three thematic sections, each reflecting key methodological and practical concerns discussed at GDN 2025:

- The first section on "**AI, Ethics, and Societal Impact in Group Decision and Negotiation**" consists of three papers. Balle et al. designed negotiation agents with an AI-based approach for teaching various negotiation styles to (novice) negotiators. Kröcher et al investigated the impact of socio-demographic factors on negotiation ethics, revealing significant insights into cultural differences, age-related trends, and the nuanced relationship between negotiation experience and ethical behavior. Chebotarev analyzed the evolution of a two-component society using the ViSE (Voting in Stochastic Environment) model, examining changes in cooperation levels (group size g) and degrees of conservatism (voting threshold t).
- The second section of this volume comprises four papers related to "**Preference Modeling, Evaluation, and Decision Support in Group Contexts**". Souza et al. proposed a Multicriteria Group Decision Making (MCGDM) model for ranking digital technologies applied to energy management problems in Brazilian industries. Lakmayer and Danielson examined how sampling alternative values from different distributions affects the performance of surrogate weight methods in additive models within multi-criteria decision analysis. Santos-Garcia and Alcantud Santos-García contributed to

the literature by building on the concept of consistent preference-approval structures and approbatory social welfare functions. Wachowicz and Roszkiwska analyzed the evaluation of the eNego electronic negotiation system, considering both subjective user feedback and objective performance metrics.

- The last section, on **"Conflict Modeling in Complex Decision Environments"** contains five studies of different strategic decision processes. Zhu, Kilgour, and Hipel presented a refinement of sequential stability concepts in multi-decision-maker conflicts, proposing more robust definitions of sanctions within the graph model paradigm to address ambiguities in existing cooperative and non-cooperative frameworks. Gu et al. presented an extension of the Graph Model for Conflict Resolution (GMCR) by incorporating triangular fuzzy preferences, offering a comprehensive framework for analyzing strategic conflicts under uncertainty with refined stability concepts. Silva, Morais, and Fang proposed integration of strength and probabilistic preferences into the GMCR, enhancing the expressiveness of decision-maker behavior in strategic conflict analysis. Xu, Maemura, and Ozawa evaluated the impacts of various transport policy schemes on the pricing dilemma to draw implications for transport policy making in Singapore and Hong Kong, highlighting the nuanced roles of fare, subsidy, and finance schemes in balancing affordability and profitability. Klan and Horita proposed a Group Support System (GSS) integrating the Building Information Modeling (BIM) and Nash bargaining to enhance dispute resolution in tunnel projects, offering a data-driven, transparent approach that improves negotiation efficiency and supports the digital transformation of construction management.

We would like to take this opportunity to express our sincere appreciation to many people for their work in organizing GDN 2025 and preparing this volume. Particularly, special thanks go to Liping Fang and Pascale Zaraté, Honorary Chairs of GDN 2025, and Ginger Ke, Program Chair of GDN 2025, for their contributions in organizing GDN 2025, and to the Group Decision and Negotiation (GDN) Section, Institute for Operations Research and the Management Sciences (INFORMS), in general. We are also thankful to all of the Stream Organizers: Liping Fang, Keith W. Hipel, and D. Marc Kilgour (Conflict Resolution); Mareike Schoop, Rudolf Vetschera, and Muhammed-Fatih Kaya (Negotiation Support Systems and Studies (NS3)); Tomasz Wachowicz and Danielle Costa Morais (Preference Modeling for GDN); Pascale Zaraté and Guy Camilleri (Collaborative Decision Making); Zhen Zhang, Yucheng Dong, Francisco Chiclana, and Enrique Herrera-Viedma (Intelligent Group Decision Making and Consensus Process); Haiyan Xu, Shawei He, and Shinan Zhao (Risk Evaluation and Negotiation Strategies); Gilberto Montibeller, Jarrod Goentzel, and Milena Janjevic (Facilitated Decision Modeling in the Codesign of Operations Systems); Dominik Siemon, Muhammed-Fatih Kaya, and Edona Elshan (Artificial Intelligence in GDN); Alberto Turón, Jorge Navarro, and José María Moreno-Jiménez (Human, Social, and Artificial Cognition in GDN).

We would also like to sincerely thank the reviewers for their informative and prompt reviews of papers. They are: Ana Paula Costa, Benshuo Yang, Eduarda Frej, Ewa Roszkowska, Ginger Ke, Haiyan Xu, Hannu Nurmi, Jing Yu, Junjie Wang, Leandro C. Rego, Lucia Roselli, Mareike Schoop, Masahide Horita, Nannan Wu, Pascale Zaraté, Rudolf Vetschera, Rustam Vahidov, Shinan Zhao, Yi Xiao, and Yuming Huang.

We are grateful to the staff at Springer for their excellent support.

April 2025

José María Moreno-Jiménez
Danielle Costa Morais
María Teresa Escobar
Alberto Turón

Organization

Honorary Chairs

Liping Fang Ryerson University, Canada
Pascale Zaraté Toulouse Capitole University, France

General Chairs

José María Moreno-Jiménez Universidad de Zaragoza, Spain
Danielle Costa Morais Federal University of Pernambuco, Brazil

Program Chairs

María Teresa Escobar Universidad de Zaragoza, Spain
Alberto Turón Universidad de Zaragoza, Spain
Ginger Ke Memorial University of Newfoundland, Canada

Program Committee

Abdelkader Adla Oran 1 University, Algeria
Adiel Teixeira de Almeida Federal University of Pernambuco, Brazil
Alexis Tsoukias Paris Dauphine University, France
Ana Paula Cabral Seixas Costa Federal University of Pernambuco, Brazil
Antonio De Nicola ENEA, Italy
Ben C. K. Ngan Worcester Polytechnic Institute, USA
Carolina Lino Martins Federal University of Mato Grosso do Sul, Brazil
Danielle Costa Morais Federal University of Pernambuco, Brazil
Edona Elshan Vrije Universiteit Amsterdam, Netherlands
Ewa Roszkowska Białystok University of Technology, Poland
Fátima Dargam SimTech Simulation Technology & REACH Innovation, Austria
François Pinet IRSTEA, France
Fran Ackermann Curtin University, Australia
Fuad Aleskerov HSE University, Russia
Ginger Ke Memorial University of Newfoundland, Canada

G.-J. de Vreede	University of South Florida, USA
Guy Camileri	IRIT, France
Haiyan Xu	University of Aeronautics and Astronautics, China
Hannu Nurmi	University of Turku, Finland
Jason Papathanasiou	University of Macedonia, Greece
John Zeleznikow	La Trobe University, Australia
José María Moreno-Jiménez	University of Zaragoza, Spain
Keith Hipel	University of Waterloo, Canada
Konstantinos Vergidis	University of Macedonia, Greece
Liping Fang	Toronto Metropolitan University, Canada
Luis Dias	University of Coimbra, Portugal
Marc Kilgour	Wilfrid Laurier University, Canada
Mareike Schoop	University of Hohenheim, Germany
María Teresa Escobar	Universidad de Zaragoza, Spain
Masahide Horita	University of Tokyo, Japan
Michael Filzmoser	TU Wien, Austria
Muhammed-Fatih Kaya	University of Hohenheim, Germany
Nannan Wu	Hong Kong Polytechnic University, China
Nikolaos Matsatsinis	Technical University of Crete, Greece
Pavlos Delias	Kavala Institute of Technology, Greece
Pascale Zaraté	Université Toulouse 1 Capitole, France
Przemyslaw Szufel	Warsaw School of Economics, Poland
Rudolf Vetschera	University of Vienna, Austria
ShiKui Wu	Lakehead University, Canada
Sandro Radovanovic	University of Belgrade, Serbia
Sean Eom	Southeast Missouri State University, USA
Shawei He	Nanjing University of Aeronautics and Astronautics, China
Tomasz Szapiro	Warsaw School of Economics, Poland
Tomasz Wachowicz	University of Economics in Katowice, Poland
Tung X. Bui	University of Hawaii, USA
Yu Maemura	University of Tokyo, Japan
Zhen Zhang	Dalian University of Technology, China

Organizing Committee

José María Moreno-Jiménez	Victoria Muerza
Juan Aguarón	Laura Muñoz
Alfredo Altuzarra	Jorge Navarro
María Teresa Escobar	Alberto Turón

Contents

AI, Ethics, and Societal Impact in Group Decision and Negotiation

AI-Based Negotiation Style Training 3
   Max Balle, Marlene Meyer, and Mareike Schoop

Reconsidering the SINS II Scale: Exploring Its Limitations
and Requirements for Future Research Designs 18
   Felix Kröcher, Peter Kesting, and Remigiusz Smolinski

Evolution of Society Caused by Collective and Individual Decisions:
A ViSE Model Study 34
   Pavel Chebotarev

Preference Modeling, Evaluation, and Decision Support in Group Contexts

Selection of Digital Technologies for Energy Management: A Group
Decision Approach Based on PROMETHEE-ROC 53
   Pedro Henrique Gouveia de Souza, Nathália Jucá Monteiro,
   Sergio Eduardo Gouvea da Costa, and Eduarda Asfora Frej

Stability of Surrogate MCDA Weights Under Different Assumptions
on Value Distributions 68
   Sebastian Lakmayer and Mats Danielson

The Approbatory Social Welfare Function: First Results 82
   Gustavo Santos-García and José Carlos R. Alcantud

Evaluating the eNego System: A Dual Perspective on Subjective
Acceptation and Objective Scoring System Accuracy 94
   Tomasz Wachowicz and Ewa Roszkowska

Conflict Modeling in Complex Decision Environments

Redefining Sequential Stability for Multi-decision-Maker Conflict Within
Graph Model 111
   Ziming Zhu, D. Marc Kilgour, and Keith W. Hipel

Triangular Fuzzy Preferences in the Graph Model with Two
Decision-Makers 124
   *Tianyang Gu, Bingfeng Ge, Wanying Wei, Sining Han, Zihui Liu,
and Chi Wang*

Strength of Preference and Probabilistic Preference in the Graph Model
for Conflict Resolution for Two Decision-Makers 136
   Elton César dos Santos Silva, Danielle Costa Morais, and Liping Fang

Balancing Affordability and Profitability in Urban Public Transport:
Implications from Singapore and Hong Kong 150
   Kai Xu, Yu Maemura, and Kazumasa Ozawa

A Group Support System for Resolving Variation Disputes in Tunnel
Construction Projects 165
   Muhammad Tajammal Khan and Masahide Horita

Author Index 181

AI, Ethics, and Societal Impact in Group Decision and Negotiation

AI-Based Negotiation Style Training

Max Balle, Marlene Meyer, and Mareike Schoop(✉)

University of Hohenheim, Schwerzstrasse 40, 70599 Stuttgart, Germany
{max.balle,marlene.meyer,schoop}@uni-hohenheim.de

Abstract. Knowing one's own negotiation style and recognising the partner's negotiation style are fundamental prerequisite to negotiate effectively. Since novice negotiators often fail to state these styles correctly, a dedicated training to support the identification of negotiation styles is required. In this paper, an AI-based training approach is presented. As a first step, software agents acting as training partners have to be designed. We implemented a genetic algorithm to shape the agents' behaviour. As a first implication, the designed agents act according to their intended negotiation styles without the need for extensive datasets of previous negotiations to specify those styles.

Keywords: Negotiation Styles · E-Training · Digital Negotiation · Agents · Genetic Algorithm

1 Introduction

Digital negotiations are present in everyday life [24]. They provide dedicated negotiation support through the use of information technology by means of negotiation support systems (NSSs). The support functionalities comprise communication support, decision support, document management, and conflict management [39].

Digital negotiations require system skills as well as negotiation skills. Therefore, dedicated training for both is required [25, 26]. If effective and efficient digital negotiations are the goal, the training has to be conducted in a digital form, e.g. among human negotiators using an NSS [42] or in a human-machine interaction via NSSs where the machine part is performed by intelligent software agent. The latter has the advantage of not relying on the availability and skills of other human negotiators and of not being dependent on moods, expertise, competencies, motivation and engagement of those [9, 22]. Negotiation agents are software that makes decisions in negotiations on behalf of a human principal, and thus, automate at least one negotiation activity, such as conducting in negotiations [16, 21]. Negotiation agents have been used extensively in the early 2000s. However, agents up to now do not consider individual negotiation styles in detail rather than providing behaviour orientations (cooperation vs. competition) nor do they apply different styles in training human negotiators [1, 5, 12, 31]. However, agents have the potential of displaying several styles and thus enabling novice negotiators to experience such behaviour, to learn how to react, and, most importantly, to recognise their own negotiation style(s). It was shown that novices have problems with all of these goals

[28]. This paper thus designs negotiation agents with an AI-based approach for teaching various negotiation styles to (novice) negotiators focusing on the decision-making process in a first step. An additional benefit of our AI approach with a decision-making focus is that it does not require the large amount of training data usually called for in AI applications.

Thus, the aim of this research contribution is to 1) design AI-agents for the decision-making process in negotiations without the need of extensive learning data, and 2) examine to what extent those agents can imitate the negotiation styles.

2 Methodological Approach

The methodology is based on a design science research (DSR) approach [13] (cf. Fig. 1).

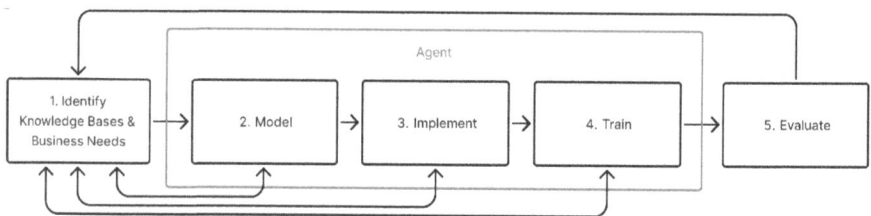

Fig. 1. Methodical Approach

First, the knowledge base and the business needs are identified. As negotiators have to be trained and since such training through an agent can be more beneficial than with a human partner [11], we examined the literature (i.e. knowledge base) for negotiation agents. Since our findings revealed that those agents do not consider negotiation theories, especially regarding negotiation styles, the aim of our research contribution (i.e. business needs) is to design and implement agents considering negotiation styles for decision-making by using an AI-based approach.

For each of the following phases (phase 2–5), the previous step is used as a basis and is adapted with new knowledge from the knowledge base if the provided information from the previous step is not sufficient.

The second phase includes the modelling of the artefact by identifying requirements of the agents, and designing a web-based system to enable human negotiators (later called participants) to negotiate with those agents. The requirements for the agents are identified by conducting a systematic literature search, which is based on an adapted four-step process as outlined by [40]. The systematic literature search has considered peer-reviewed articles in English only, with the publication date ranging from 2017 to the present. The search strings encompass agents, training, and negotiation.

In order to implement the agents in phase 3, an AI model has to be chosen. As neither a fitting pre-trained model nor a sufficiently large dataset of previous negotiations are available, these conditions have to be considered in the selection of the appropriate AI model for our agents.

In the training phase, the implemented agents are trained through conducting multiple negotiations with other agents. A detailed description of the training is presented in Sect. 5.

In the last phase, an experiment is conducted with twenty-seven participants who negotiate with the agent and fill in a survey.

3 Theoretical Background

3.1 Negotiation Styles

Negotiation styles are orientations how someone can behave in a negotiation. The orientations are based on how strong the concerns about one's own goals and the relationship to the partner are pronounced. Regarding this, five types of negotiation styles are derived: accommodating when the relationship to the partner is most important and is shown by yielding to the partner's preferences; competing when the own outcome is the most important; compromising when both interests are of concern and a compromise is attempted, e.g., by splitting the differences; collaborating when a win-win result satisfying both parties is tried to be achieved, and avoiding when neither the relationship nor the own outcome are relevant; [23].

3.2 Artificial Intelligence

There are two dimensions of AI to consider 1) measurement of intelligence in AI (is the intelligence of AI measured in comparison to human intellect or with respect to rationality?) [33], and 2) quality type of intelligence (do we search for intelligence as an internal quality or externally in the form of intelligent behaviour?) [27, 33]. Rationality refers to choosing the right action with respect to a certain goal and considering resource limitations [14]. Based on these two dimensions, four different perspectives on the topic are derived, namely acting humanly, thinking humanly, thinking rationally, and acting rationally [33] (cf. Fig. 2).

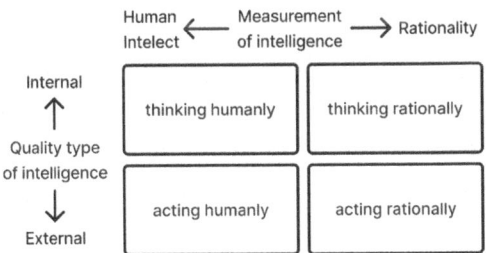

Fig. 2. Perspectives on AI (adapted from [33])

Amongst the current AI research, the perspective *acting rationally* is the most dominant approach [33]. One of these branches of AI is Machine Learning (ML). As a scientific field, ML provides computers with the ability to learn without explicit programming [35]. ML is programming computers "to learn from experience [to] eliminate the need for much [...] programming effort" [35]. A precise definition of ML programs is given by Mitchell: "A computer program is said to learn from experience E with respect to some class of tasks T and performance measure P, if its performance at tasks in T, as measured by P, improves with experience E" [30].

The three main types of ML are supervised, unsupervised, and reinforcement learning. Supervised learning trains the AI with labelled data – pairs of data with input and expected corresponding output – while unsupervised learning uses unlabelled data, i.e., without receiving explicit feedback [34]. Reinforcement learning allows the agent to learn from its actions by receiving feedback in form of rewards or punishments [34].

There are also emergent models of learning, such as the genetic algorithm, which is inspired by evolution [27]. Genetic algorithms (GAs) are characterised by evolution through mating and mutation [15]. GAs work by first initialising a population of potential solutions, then the population is repeatedly evaluated, reproduced, and replaced. The evaluation is based on evaluating the fitness of the individuals from the population for the respective task. During the reproduction, descendants are generated from the population by using genetic operators. Finally, the descendants are used to replace a part – mostly the unfit individuals – of the previous population [27]. Multiple strategies exists for the selection process and for the genetic operators recombination and mutation [20, 29].

Deep Learning is a specific type of machine learning, which is characterised by neural networks with many layers. Artificial Neural Networks (ANNs) are inspired by neurons [27] and can represent complex non-linear functions with computational circuits [34]. ANNs are called feedforward neural networks, if the output of each unit in the network only connects to the next layers. ANNs are defined as recurrent neural networks (RNNs), if units can cycle their outputs back into previous layers [34]. This cycling results in RNNs being able to memorise, thus, making them particularly useful with sequential data [33]. Gated Recurrent Units (GRU) are a type of RNNs proposed by Cho et al. in 2014 [4] offering improved memory length and reduced the risk of vanishing or exploding gradients [6].

4 Designed Artefact

In this chapter, the artefact for negotiation style training with AI-based agents is designed including the derived requirements for the artefact, a detailed description of the client-server-architecture of the system, and the designed agents.

4.1 Requirements

The systematic literature review was conducted using search strings including the terms "agents", "training", and "negotiation" in order to derive requirements for the intended agents. The review yielded six requirements, which are listed in descending order according to the frequency with which they were identified in the literature search. The first requirement is generally the most relevant to implement for our purpose. However, depending on the focus of the application of the agents, the order might change.

1. *Realistic*: Most importantly, agents have to behave realistically. This is supposed to ensure transferability of the learning to real-world human-to-human negotiation. [7, 41, 43].
2. *First-order Theory of Mind (ToM)*: This is the second most common requirement. To negotiate realistically and integratively, the agents must be capable of reasoning over beliefs and intentions the other side holds – especially regarding the utility values. [3, 17, 36, 41].
3. *Feedback*: For trainees, appropriate feedback is of high importance when learning the skills that comprise negotiation [7, 18, 44].
4. *Share information*: Agents should be able to implicitly and/or explicitly share information, mostly about their preferences. The negotiation partners rely on this information to perform ToM themselves and overcome the fixed-pie-bias [17, 36].
5. *No knowledge about the partner's utility*: Just like in most real-life negotiations, agents should have no prior knowledge of their partner's utility values. This requirement is also the reason agents have to perform ToM and engage in information sharing. [19, 32].
6. *Use history of exchange*: Finally, the agents have to be able to factor all prior exchanged messages into their decision on the next move. Again, this is necessary for realistic behaviour and to perform ToM [37, 43].

Regarding the quality type of intelligence (cf. Fig. 2), all of the requirements contain the external view on how the agent directly interacts with participants. Requirements 2, 5 and 6 also consider the internal view on the agent's decision-making process.

4.2 Client-Server Architecture of the Web-Based System

The focus of this work is on designing agents as realistic negotiation partners (cf. Sect. 4.1) that support negotiation training, hence a corresponding client-server-architecture for negotiation training is designed (cf. Fig. 3).

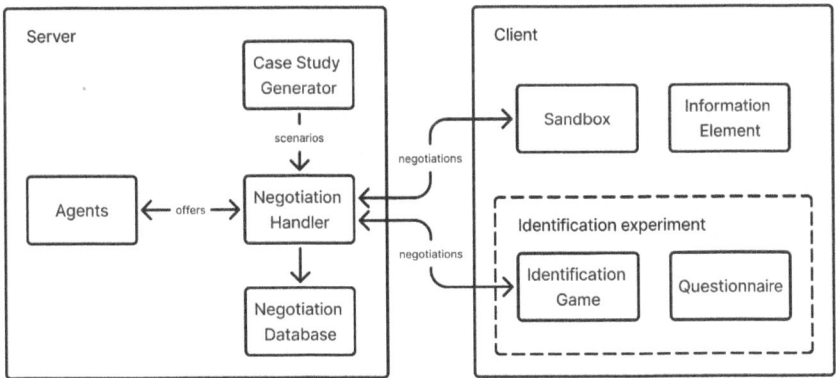

Fig. 3. Architecture

The client provides the user interface to enable the participants to negotiate with the agents either in a sandbox or during an experiment. Further, the interface provides fundamentals of negotiation theory and the user interface to ensure a uniform understanding of terms. The server contains the agents, a case study generator, and the negotiation handler with a database. The negotiation handler manages the initialisation of a new negotiation by generating a case study, and all states of the ongoing negotiations, i.e., all (counter-)offers between the negotiators (agent and participant) are managed though the negotiation handler. The case study generator and the agents are further specified in the following chapters.

4.3 Case Study Generator

The usage of a case study generator which produces new collaborative case studies with preference models for each negotiation ensures that the participants do not have a learning effect through repetitive case studies when negotiating multiple times. All case studies generated for this work contain five issues and five different options for each issue.

The relative importance of the issues for each preference model are generated independently of the partner's preference model by iteratively expanding the number of issues, starting with a set containing one issue with a relative importance of 100%. Then, the greatest value is split into two randomly sized parts with a centred triangular distribution. The calculation stops when the set contains the defined number of issues for the preference model and the issues get shuffled. The aim of this calculation is to reduce the likelihood of irrelevant issues (with a relative importance that is too low) compared to models generated by, e.g. simply splitting 100% at n-1 points.

For each issue, the preferences of each issue selection are generated considering the partner's preference model to generate actual disputes in the case study. For each issue in each preference model, the preferred option (100%) is selected at random. For both preference models, a triangular distribution is used, but one is left-sided and the other right-sided. The remaining issue selections are filled in based on the distance to the preferred option of the issue and altered by Gaussian noise.

To ensure a generated case study fulfils the requirements, their usage is verified through the availability of a fair outcome and the analysis of the pareto frontier, i.e., a pareto-optimal outcome exists with a contract imbalance of $<= 20\%$, and the pareto frontier is curved enough (extremes must be at least 0.5 apart and the area between the pareto frontier and a line connecting the extremes must be above 0.04 on the pareto graph). Thus, only case studies fulfilling the requirement are applied; otherwise, the generation process restarts.

4.4 Designed AI-Based Agents

When choosing an AI-method to be applied as a negotiation partner, the negotiation domain and the limited availability of suitable training data have to be considered besides further requirements (cf. Sect. 4.1). Characteristics of negotiations, such as its sequential nature and the dependencies on the negotiation partner regarding the outcome of the negotiation, make negotiation a complex domain [2, 38]. Based on the previous specified challenges of negotiating with a realistic agent, we use a Deep Learning approach. Following the ML definition by [30], for each negotiation style S we define the task T as "Negotiate like a human with S" and measure performance P with a function mapping negotiation result and S to a fitness score. To provide the experience E, negotiation simulations between the agents are conducted.

Deep Learning is realised by applying a Genetic Algorithm to RNNs. This utilises the abilities of RNNs to model complex non-linear functions and to remember past input [34] and combines it with GAs easy parallelisation and balance of exploration and exploitation [15, 20, 29].

The recurrent layers of the model are GRU-based and supplemented with dense, non-recurrent layers (cf. Fig. 4).

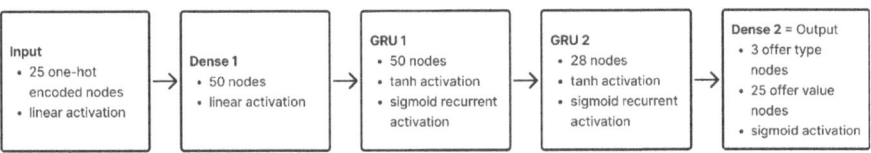

Fig. 4. RNN layers

There are fifty inputs for each new offer, two for each option in each issue: the utility of this option as the product of importance and preference, and in One-Hot-encoding, whether it is chosen. One-Hot-encoding specifies that for every issue, only the chosen option is encoded as "1", the others as "0" [10]. The model output consists of 28 nodes, a group of three for each possible message types (counteroffer/accept/reject) and five groups of five for the issues. In each group, the option corresponding to the node with the highest output is the choice of the model for its new offer.

Previous offers influence the outcome via the recurrence. Using two GRU layers allows for a recurrence along the reduction from 50 to 28 hidden units.

The performance P of an agent in a negotiation is measured with a fitness function. The fitness function varies for each negotiation style and calculates a fitness score for a negotiation outcome by calculating the weighed sum of these five factors (cf. Table 1): own utility, partner's utility, joint utility, fairness, and time. The negotiator's own utility and the partner's utility are directly based on the dimensions of the negotiation style (own concern vs. partner's concern). Joint utility is defined to measure the extent of collaborating by adding up own utility and partner's utility. Fairness specifies the contract imbalance between the own utility and the partner's utility (absolute difference). Time is considered by the additive inverse of the normalised number of exchanged messages and reflects how urgent ending or continuing the negotiation is to the agent. The time factor can change during the training progress to reduce indecisiveness and encourage agents to take more or less time while exploring solutions.

As accommodating agents prioritise their partner's utility, they disregard their own utility, joint utility, and fairness. Competing agents, as opposite to accommodating agents, prioritise their own utility and disregard their partner's utility, joint utility, and fairness. Avoiding agents try to withdraw from the negotiation by concluding as fast as possible with little care for the outcome. Collaborating agents try to completely fulfil their own and their partners' goals simultaneously, aiming for an outcome of high joint utility and fairness. Compromising agents are aiming for a partial fulfilment of their own and the partner's goals with a more or less fair outcome.

Table 1. Fitness weights

	Time	Own utility	Partner utility	Joint utility	Fairness
Accommodating	0.25/−0.025/0	0	1	0	0
Collaborating	0.1/−0.01/0	1	1	1	1
Compromising	0.75/−0.075/0	0.25	0.25	0.5	1
Avoiding	1	0	0	0	0.25
Competing	0.25/−0.025/0	1	0	0	−1

5 Training of the Agents

The designed AI-based agents have to learn how to act according to their negotiation styles through training. As a training algorithm we developed a genetic algorithm inspired by [15, 20, 29].

Due to different characteristics of each negotiation style, a different population is considered for each negotiation style in the training algorithm. Initially, there are 100 individuals per population – rising to 500 in late generations – and every agent is initialised with random weights. The genome of each individual consists of the weight matrices of its RNN. This makes this approach a real-coded GA [20]. For each generation, four consecutive steps of the GA – Negotiation simulation, selection, recombination, and mutation – are executed (cf. Fig. 5).

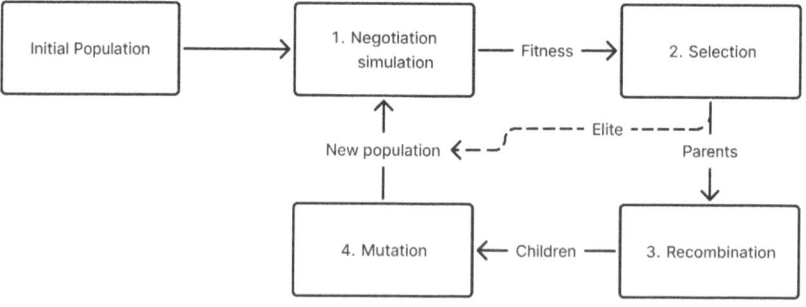

Fig. 5. Steps of the GA

After the population is initialised, the fitness of every agent is determined by a negotiation simulation (step 1). To conduct one turn of the negotiation simulation, a case study has to be used (generated by the case study generator) and the agents have to negotiate with each other. More precisely, each agent negotiates twice with one random agent of each negotiation style (in total 10 negotiations per turn) to allow switching roles and compensating unfairness in a case study. The negotiation length is limited to the exchange of 100 messages each during the negotiation to ensure termination even with indecisive models. In early generations, the negotiation simulation of one generation consists of one turn. Later generations have up to three turns to achieve higher accuracy of the estimation. The outcomes of all negotiations of an agent are judged with the fitness function under the agent's negotiation style (cf. Sect. 4.4) and the fitness scores consolidated into this agent's fitness.

Once every agent has a determined fitness, the selection (step 2) is conducted. To do so, a fixed part of the population dies without reproduction. Some approaches randomise the selection with odds favouring the fit individuals [20]. We apply a simple cut-off of the unfit 50% since step 1 already introduced chances of survival for weaker individuals when giving them easy pairings. The remaining individuals reproduce. To ensure fit individuals produce more descendants, this GA takes a staggered approach where subgroup reproduce amongst themselves; first the top 10%; followed by the top 30% and finally the top 50% (proportions taken from early trials). An elitist selection is applied to the descendants as the best individuals (top 10%) directly survive into the next generation without modification [29]. This is different to evolution in nature, but inspired by other GA approaches where the best individual survives. The benefit of elitist selection lies in the prevention of loss of development [29]. Because of luck involved in determining the fitness during step 1, more elite individuals survive in our GA, which slows down the evolution, but increases the likelihood of keeping the best agent. In each generation the same number of agents die as new agents are introduced, which keeps the size of the population consistent.

Reproduction happens in pairs of two, i.e., two parent agents combine to produce two children. During the recombination (or crossover) each of the different RNNs weight matrices of the parents' genomes get extracted and split. The split happens at a random point along on a random dimension of the matrix. Corresponding matrices of the two parents split at the same point. The weight matrices of both children are then the different combinations of the parts of the corresponding parent's matrices. This is a special case of multi-point crossover with regards to the whole genome [8] and is illustrated in Fig. 6.

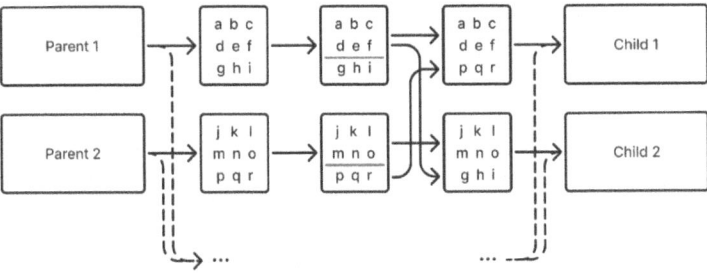

Fig. 6. Exemplary recombination of one matrix of the genome

Each child is then individually mutated, which is realised by Gaussian noise to each weight, and thus, increases the variance in the population.

When the steps of the GA terminate, the fittest individual of each of the five populations corresponding to the negotiation styles is chosen as the final model for its style by repeating step 1 of Fig. 5 with the final population to increase the likelihood of choosing the best model.

6 Results of the Experiment

In order to validate the generated agent, the defined client-server-architecture was implemented and an experiment was conducted. This chapter includes the results of the conducted experiment 1) by identifying the negotiation styles of the agents and 2) by evaluating the usefulness of the agents.

There were 27 participants in the experiment (12 female, 15 male). 21 of the participants are between 18 and 24 years old; 3 participants are between 25 and 34 years; one participant is between 45 and 54 years; and two participants are between 55 and 64 years old. 21 participants have an upper secondary education, the remaining participants have a bachelor's degree or above. 13 participants disclosed to having no prior negotiation experience; 11 participants claimed to have some experience, and three participants reported to be (highly) experienced.

In the first part of the experiment, the participants were asked to negotiate and then indicate the negotiation style they associated with their negotiation partner (aka implemented agent). The participants were asked to fill in a short survey to share some information about their demographic data and their negotiation experience. Further, they received all relevant information about the setting of the negotiation including the case study. Every case study used during the experiment was produced by our case study generator and only used once.

In each negotiation they conducted, the participants were paired with an agent of a specific negotiation style which was unknown to the participants. After completion of the negotiation, the participants were asked to indicate the negotiation style of their partner of the current negotiation. The participants were allowed to conclude as many negotiations as they like. However, none negotiated with more than two agents with the same style.

This led to 52 data points in total. 14 times the participants were able to state the negotiation style of their partner correctly. The results are differentiated between those participants with no experience (No Exp.) and those with some or more experience (Exp.) (cf. Table 2). Six participants (No Exp. 3 of 6 participants; Exp. 3 of 5 participants) identified the competing style correctly. None identified the collaborating style (No Exp. 4 participants, Exp. 2 participants) and the compromising style (No Exp. 10 participants, Exp. 2 participants) correctly. Avoiding was identified correctly five times (No Exp. 1 of 6 participants, Exp. 4 of 6 participants). Three participants identified the accommodating style correctly (No Exp. 2 of 6 participants, Exp. 1 of 4 participants).

After the participants concluded the negotiations, they were asked to complete a post-negotiation survey by rating statements on a 7-point Likert scale regarding the usefulness of the system and how realistic the agents proceeded (cf. Table 3). The conducted system

Table 2. Style Identification Differentiated by the Participants based on their Experience

Judgments		Agent				
		Competing	Collaborating	Compromising	Avoiding	Accommodating
Competing	No Exp	3	0	0	0	2
	Exp	3	0	0	0	2
Collaborating	No Exp	1	0	3	0	2
	Exp	0	0	0	0	1
Compromising	No Exp	1	0	0	2	0
	Exp	2	0	0	1	1
Avoiding	No Exp	0	2	5	1	0
	Exp	0	1	1	4	0
Accommodating	No Exp	1	2	2	3	2
	Exp	0	1	1	1	1
Total	No Exp	6	4	10	6	6
	Exp	5	2	2	6	4

Table 3. Likert statements and reactions

Statement	Mean	SD
This tool is helpful when learning about different negotiation styles	5.48	1.32
This tool is helpful to train the identification of the negotiation opponents' styles	5.38	1.20
The AI models behaved realistic	4.10	1.55
The AI models seem to work with a model of my beliefs and intentions	4.90	1.14

tend to support the identification of the partner's negotiation style (mean = 5.38, SD = 1.20) and the learning of different negotiation styles (mean = 5.48, SD = 1.32). The AI-models behaving realistic is rated as neutral by the participants (mean = 4.10, SD = 1.55). The participants tend to agree that the AI models seem to work with a model of their beliefs and intentions (mean = 4.90, SD = 1.14). 21 of the participants completed the survey (9 females; 12 males), 16 of whom were between 18 and 24 years old; two were between 25–34 years old; one was between 45–54 years old; and two were between 55–64 years old.

7 Concluding Discussion

Since the sample size of the conducted experiment is not very large, further investigations have to be conducted. However, the results already show some first implications.

The negotiation styles were correctly identified by 14 participations (26.92%) which is higher than the expected 20% of unsubstantiated decisions. However, their identification varies highly between the styles; Styles in which the agents do not focus on the relationship at all (namely competing and avoiding), were identified correctly in a higher number than the remaining styles; competing with 54.55% (6 of 11), avoiding with 41.67% (5 of 12), accommodating with 27.27% (3 out 11), compromising with 0%, and collaborating with 0%. Since the agents currently only consider the utility values, it can be more challenging to identify them correctly as the utility value is the only parameter to evaluate whether the agents are concerned with the relationship or not. The used preference models in the negotiations are indicated to be collaborative, i.e., the agent could fulfil its behaviour accurately by making concessions in terms of the relationship, but those concession could have a negative effect on the participant's view. Thus, further investigations have to be conducted to evaluate whether the compromising and collaborating agents behave according to the style definitions, e.g., with a larger number of sample size and/or with textual offer exchanges to underline the relationship aspect.

Further, participants with at least some experience were able to identify the models with 40.00% accuracy in general and, thus, performed far better than participants without experience, who only identified correctly with a rate of 18.75%. Thus, negotiation experience supports the correct identification of the partner's style.

The results of the post-negotiation survey (see Table 3) indicate that the approach taken by this work in teaching negotiation styles is well received both as a tool to learn about negotiation styles as well as in identifying them. Thus, the requirement *realistic* is fulfilled. Further, the participants tend to agree with the sentiment that the agents work with a model of their beliefs and intentions, i.e., perform *ToM*. The participants rated the AI-models realistic behaviour as neutral which can be caused by not providing explanations to the offers. The results of Table 3 correspond to the first two derived requirements (cf. Sect. 4.1). The requirement *giving feedback* was fulfilled by revealing the imitated negotiation style after participants gave their judgements, allowing them to reflect on their identification. The agents *shared information* implicitly via the offers values as they are non-textual. The remaining requirements (*no knowledge about the partner's utility* and *use history of exchange*) are met by the design of the agents.

There are some limitations of this research contribution. The systematic literature review is limited by the search string and the data. As already mentioned, the sample size is currently small and further experiments have to be conducted. However, the first implications indicate that the designed agents work according to their purpose and mostly imitate their negotiation styles correctly for the decision-making process. This research contribution demonstrates that no extensive datasets are needed to build agents which can negotiate with specific negotiation styles rather than genetic algorithms can be implemented and trained to do so. However, it is important to note that the agents' current focus is on decision-making. In future research, it would be beneficial to consider the influence of additional factors on real-world negotiations, such as communication and emotions. These factors warrant further examination to enhance our understanding of the complex dynamics involved in negotiations.

As a next step, the agents should be validated in a further experiment by increasing the sample size and conducting a statistical test. Depending on the results, the agents' negotiation styles could be refined though alterations to the genetic algorithm and the incentives (fitness and corresponding weights) or could be compared to different AI methods.

References

1. Baarslag, T., Hendrikx, M.J.C., Hindriks, K.V., Jonker, C.M.: Learning about the opponent in automated bilateral negotiation: a comprehensive survey of opponent modeling techniques. Auton. Agent. Multi-Agent Syst. **30**(5), 849–898 (2016). https://doi.org/10.1007/s10458-015-9309-1
2. Bichler, M., Kersten, G., Strecker, S.: Towards a structured design of electronic negotiations. Group Decis. Negot. **12**(4), 311–335 (2003). https://doi.org/10.1023/A:1024867820235
3. Bouman, K., Lefter, I., Rook, L., Oertel, C., Jonker, C., Brazier, F.: The need for a female perspective in designing agent-based negotiation support. In: Martinho, C., Dias, J., Campos, J., Heylen, D. (eds.) Proceedings of the 22nd ACM International Conference on Intelligent Virtual Agents, pp. 1–8. ACM, New York (2022). https://doi.org/10.1145/3514197.3549691
4. Cho, K., van Merrienboer, B., Gulcehre, C., Bahdanau, D., Bougares, F., Schwenk, H., et al.: Learning phrase representations using RNN encoder-decoder for statistical machine translation. In: Moschitti, A., Pang, B., Daelemans, W. (eds.) Proceedings of the 2014 Conference on Empirical Methods in Natural Language Processing (EMNLP), pp. 1724–1734. Association for Computational Linguistics, Doha (2014). https://doi.org/10.48550/arXiv.1406.1078
5. Choi, S.P., Liu, J., Chan, S.-P.: A genetic agent-based negotiation system. Comput. Netw. **37**(2), 195–204 (2001). https://doi.org/10.1016/S1389-1286(01)00215-8
6. Dey, R., Salem, F.M.: Gate-variants of gated recurrent unit (gru) neural networks. In: IEEE (ed.) 60th International Midwest Symposium on Circuits and Systems (MWSCAS), pp. 1597–1600 (2017). https://doi.org/10.1109/MWSCAS.2017.8053243
7. Dinnar, S., Dede, C., Johnson, E., Straub, C., Korjus, K.: Artificial intelligence and technology in teaching negotiation. Negotiat. J. **37**(1), 65–82 (2021). https://doi.org/10.1111/nejo.12351
8. Eshelman, L.J., Caruana, R., Schaffer, J.D.: Biases in the crossover landscape. In: Schaffer, J.D. (ed.) Proceedings of the 3rd International Conference on Genetic Algorithms, pp. 10–19 (1989)
9. Forgas, J.P.: On feeling good and getting your way: mood effects on negotiator cognition and bargaining strategies. J. Pers. Soc. Psychol. **74**(3), 565–577 (1998). https://doi.org/10.1037/0022-3514.74.3.565
10. Goodfellow, I., Bengio, Y., Courville, A.: Deep Learning. MIT Press, Cambridge (2016)
11. Gratch, J., DeVault, D., Lucas, G.: The benefits of virtual humans for teaching negotiation. In: Traum, D., Swartout, W., Khooshabeh, P., Kopp, S., Scherer, S., Leuski, A. (eds.) Intelligent Virtual Agents, pp. 283–294. Springer, Cham (2016). https://doi.org/10.1007/978-3-319-47665-0_25
12. Gratch, J., DeVault, D., Lucas, G.M., Marsella, S.: Negotiation as a challenge problem for virtual humans. In: Brinkman, W.-P., Broekens, J., Heylen, D. (eds.) Intelligent Virtual Agents: 15th International Conference, IVA 2015, Delft, The Netherlands, August 26-28, 2015, Proceedings, pp. 201–215. Springer, Cham (2015). https://doi.org/10.1007/978-3-319-21996-7_21
13. Hevner, M., Park, R.: Design science in information systems research. MIS Q. **28**(1), 75 (2004). https://doi.org/10.2307/25148625

14. High-Level Expert Group on Artificial Intelligence: A definition of AI: Main capabilities and scientific disciplines (2018)
15. Holland, J.H.: Genetic algorithms. Sci. Am. **267**(1), 66–73 (1992)
16. Jennings, N.R., Faratin, P., Lomuscio, A.R., Parsons, S., Wooldridge, M.J., Sierra, C.: Automated negotiation: prospects, methods and challenges. Group Decis. Negot. **10**(2), 199–215 (2001). https://doi.org/10.1023/A:1008746126376
17. Johnson, E., Gratch, J.: The impact of implicit information exchange in human-agent negotiations. In: Proceedings of the 20th ACM International Conference on Intelligent Virtual Agents, pp. 1–8. ACM, New York (2020). https://doi.org/10.1145/3383652.3423895
18. Johnson, E., Gratch, J., Devault, D.: Towards an autonomous agent that provides automated feedback on students' negotiation skills. In: Larson, K., Winikoff, M., Das, S., Durfee, E. (eds.) Proceedings of the 16th Conference on Autonomous Agents and Multiagents Systems, pp. 410–418 (2017)
19. Johnson, E., Gratch, J., Gil, Y.: Virtual agent approach for teaching the collaborative problem solving skill of negotiation. In: Wang, N., Rebolledo-Mendez, G., Dimitrova, V., Matsuda, N., Santos, O.C. (eds.) Artificial Intelligence in Education. Posters and Late Breaking Results, Workshops and Tutorials, Industry and Innovation Tracks, Practitioners, Doctoral Consortium and Blue Sky, vol. 1831. Communications in Computer and Information Science, pp. 530–535. Springer, Cham (2023). https://doi.org/10.1007/978-3-031-36336-8_82
20. Katoch, S., Chauhan, S.S., Kumar, V.: A review on genetic algorithm: past, present, and future. Multimedia Tools Appl. **80**(5), 8091–8126 (2021). https://doi.org/10.1007/s11042-020-10139-6
21. Kersten, G.E., Lai, H.: Negotiation support and e-negotiation systems: an overview. Group Decis. Negot. **16**(6), 553–586 (2007). https://doi.org/10.1007/s10726-007-9095-5
22. Lewicki, R.J., Barry, B., Saunders, D.M.: Negotiation, 6th edn. McGraw-Hill/Irwin, Boston (2010)
23. Lewicki, R.J., Saunders, D.M., Barry, B.: Negotiation. Readings, Exercises, and Cases, 6th edn. McGraw-Hill, New York (2007)
24. Lin, R., Kraus, S., Wilkenfeld, J., Barry, J.: Negotiating with bounded rational agents in environments with incomplete information using an automated agent. Artif. Intell. **172**(6–7), 823–851 (2008). https://doi.org/10.1016/j.artint.2007.09.007
25. Loewenstein, J., Thompson, L.L.: The challenge of learning. Negot. J. **16**(4), 399–408 (2000). https://doi.org/10.1023/A:1026692922914
26. Loewenstein, J., Thompson, L.L.: Learning to negotiate: novice and experienced negotiators. In: Thompson, L.L. (ed.) Negotiation Theory and Research, pp. 77–97. Psychology Press (2006)
27. Luger, G.F.: Artificial Intelligence. Structures and Strategies for Complex Problem Solving, 6th edn. Pearson, Boston (2009)
28. Meyer, M.: An Explorative study of the usage of negotiation styles in higher education. In: UK Academy for Information Systems Conference (2022)
29. Mirjalili, S.: Genetic algorithm. In: Mirjalili, S. (ed.) Evolutionary Algorithms and Neural Networks, vol. 780. Studies in Computational Intelligence, pp. 43–55. Springer, Cham (2019). https://doi.org/10.1007/978-3-319-93025-1_4
30. Mitchell, T.M.: Machine learning. McGraw-Hill international editions computer science series. McGraw-Hill, New York (1997)
31. Mollick, E., Mollick, L.: How to use AI to create role-play scenarios for your students. Here's a sample prompt you can customize for your class (2024)
32. Petukhova, V., Sharifullaeva, F., Klakow, D.: Modelling shared decision making in medical negotiations: interactive training with cognitive agents. In: Baldoni, M., Dastani, M., Liao,

B., Sakurai, Y., Zalila Wenkstern, R. (eds.) PRIMA 2019: Principles and Practice of Multi-Agent Systems. Lecture Notes in Computer Science, vol. 11873, pp. 251–270. Springer, Cham (2019). https://doi.org/10.1007/978-3-030-33792-6_16
33. Russell, S.J., Norvig, P.: Artificial intelligence: A Modern Approach. Pearson series in artificial intelligence. Pearson, Harlow (2022)
34. Russell, S.J., Norvig, P.: Künstliche Intelligenz. Ein moderner Ansatz, 4th edn. Pearson, München (2023)
35. Samuel, A.L.: Some studies in machine learning using the game of checkers. IBM J. Res. Dev. **3**(3), 210–229 (1959). https://doi.org/10.1147/rd.33.0210
36. Sato, M., Terada, K., Gratch, J.: Teaching reverse appraisal to improve negotiation skills. IEEE Trans. Affective Comput. **15**(3), 1–14 (2024). https://doi.org/10.1109/TAFFC.2023.3285931
37. Schmid, A., Kronberger, O., Vonderach, N., Schoop, M.: Are you for real? A negotiation bot for electronic negotiations. In: UK Academy for Information Systems Conference Proceedings 2021, p. 3 (2021)
38. Schoop, M.: Support of complex electronic negotiations. In: Shakun, M.F., Kilgour, D.M., Eden, C. (eds.) Handbook of Group Decision and Negotiation, vol. 4. Advances in Group Decision and Negotiation, pp. 409–423. Springer, Dordrecht (2010). https://doi.org/10.1007/978-90-481-9097-3_24
39. Schoop, M.: Negoisst: Complex digital negotiation support. In: Kilgour, D.M., Eden, C. (eds.) Handbook of Group Decision and Negotiation, pp. 1149–1167. Springer, Cham (2021). https://doi.org/10.1007/978-3-030-49629-6_24
40. Snyder, H.: Literature review as a research methodology: an overview and guidelines. J. Bus. Res. **104**, 333–339 (2019). https://doi.org/10.1016/j.jbusres.2019.07.039
41. Stevens, C.A., Daamen, J., Gaudrain, E., Renkema, T., Top, J.D., Cnossen, F., et al.: Using cognitive agents to train negotiation skills. Front. Psychol. **9**, 154 (2018). https://doi.org/10.3389/fpsyg.2018.00154
42. Thompson, L.L.: The Mind and Heart of the Negotiator, 7th edn. Pearson, bOSTON (2022)
43. Xu, Y., Sequeira, P., Marsella, S.: Towards modeling agent negotiators by analyzing human negotiation behavior. In: IEEE (ed.) 2017 Seventh International Conference on Affective Computing and Intelligent Interaction (ACII), pp. 58–64. IEEE (2017). https://doi.org/10.1109/ACII.2017.8273579
44. Zahn, E.-M., Schöbel, S.: Let's chat to negotiate: designing a conversational agent for learning negotiation skills. In: Mandviwalla, M., Söllner, M., Tuunanen, T. (eds.) Design Science Research for a Resilient Future, pp. 229–243. Springer, Cham (2024). https://doi.org/10.1007/978-3-031-61175-9_16

Reconsidering the SINS II Scale: Exploring Its Limitations and Requirements for Future Research Designs

Felix Kröcher[1]( ), Peter Kesting[2], and Remigiusz Smolinski[1]

[1] HHL Leipzig Graduate School of Management, Jahnallee 59, 04109 Leipzig, Germany
felix.kroecher@hhl.de

[2] Department of Management, Aarhus School of Business and Social Sciences, Aarhus University, Fuglesangs Alle 4, 8210 Aarhus, Denmark

Abstract. Ethics remains a critical area of negotiation research, yet existing measurement tools in this field have unexamined limitations. This study critically analyzes the Self-reported Inappropriate Negotiation Strategies (SINS II) scale, as the most prominent instrument for assessing Ethical Ambiguous Negotiation Tactics (EANT). By examining 12,375 data points from 495 respondents, the research reveals significant challenges in the scale's application. While confirming the scale's statistical reliability, the study uncovers key methodological concerns, including content validity issues, potential social desirability bias, and the need to account for participants' missing familiarity with specific tactics. The analysis explores how socio-demographic factors influence negotiation ethics, with notable findings on cultural differences, age effects, and the complex relationship between negotiation experience and ethical behavior. The research provides crucial insights for future negotiation ethics studies, highlighting the importance of nuanced, contextually sensitive approaches to measuring ethical decision-making in negotiations. These findings offer valuable implications for negotiation training, cross-cultural research, and understanding the subtle dynamics of ethical behavior in business interactions.

Keywords: SINS II scale · negotiation ethics · behavioral ethics · ethically ambiguous negotiation tactics · social desirability bias

1 Introduction

Numerous studies highlight the prevalence and significance of unethical behavior in negotiations (e.g., Aquino and Becker 2005; Boles et al. 2000; Olekalns et al. 2014; Rottenburger and Kaufmann 2020; Schweitzer et al. 2002; Schweitzer and Croson 1999; Tenbrunsel 1998). Negotiation, defined as "back and forth communication designed to reach an agreement" (Fisher et al. 1991, p. xvii), involves ethics as a core element of the interactive, joint decision-making process. Unethical behavior in negotiations manifests in three primary ways: gaining an illegitimate advantage with potential legal consequences, causing disproportionate harm or damage, and obstructing cooperation and problem-solving, thereby limiting the realization of mutually beneficial outcomes (Banas and Parks 2002; Friedman and Shapiro 1995; Gino et al. 2010; McCornack and Levine 1990; Schweitzer et al. 2005; Shell 1991).

In quantitative research, unethical negotiation behavior is often conceptualized using the Self-reported Inappropriate Negotiation Strategies (SINS II) Scale (Robinson et al. 2000; Fulmer et al. 2009) to assess Ethical Ambiguous Negotiation Tactics (EANT). Despite its widespread application, our study identifies four primary limitations in the SINS II scale that have not been fully addressed in prior research. First, the scale shows issues regarding the prevalence of certain tactics, which may not adequately allow to measure negotiators' perception of these tactics. Second, the scale does not account for respondents' familiarity with the specific negotiation tactics, potentially biasing self-reports. Third, the measure is susceptible to social desirability bias, which can distort responses. Fourth, aggregating effects by the seven tactic factors or the total questionnaire results, as often done in prior studies, risks obscuring nuanced interactions between socio-demographic factors and individual tactics. This limitation underscores the need for a more granular analysis when applying the scale.

While the SINS II scale is reconfirmed as a statistically reliable construct, these identified limitations underscore the need for a more nuanced approach when employing the scale in future research. In response, our investigation not only tests the proposed hypotheses using 12,375 data points collected from 495 respondents but also explores among others the influence of cultural affiliation, age, and negotiation experience on the frequency and perception of ethically ambiguous tactics.

The remainder of the paper is organized as follows. In the next section, we review the relevant literature and develop our research hypotheses. Section 3 details the research methodology, including our data collection and analysis procedures. Section 4 presents the empirical results, featuring both parametric and non-parametric analyses, and discusses the identified limitations of the SINS II scale in depth. In Sect. 5, we interpret our findings in light of the existing literature and consider their implications for both theory and practice. Finally, Sect. 6 concludes the paper by summarizing the key contributions, addressing the study's limitations, and suggesting avenues for future research.

2 Literature Review and Research Hypotheses

Interest in negotiation ethics has grown significantly since the 1980s (e.g., Carson et al. 1982; Michelman 1983), evolving from early conceptual discussions (Lax and Sebenius 1986; Dees and Cramton 1991) to more empirically grounded investigations (Lewicki et al. 2021). To facilitate empirical research the SINS II scale, developed by Robinson et al. (2000) and refined by Fulmer et al. (2009), has become the most widely used tool for research on negotiation ethics. It comprises 25 items considered to be ethically ambiguous negotiation tactics, grouped into seven tactic factors. Since 1997, the scale and its predecessors have been applied in 53 quantitative studies, making it a standardized instrument for analyzing negotiation ethics. In those studies, as well as for this paper the terms SINS II scale items, referring to the tactics investigated by the SINS II scale and EANT referring to the general category of such ambiguous negotiation tactics, have been used synonymously. In addition to measuring the perceived appropriateness of the inquired tactics, the scale has also been employed to assess participants' likelihood of applying these tactics (Volkema et al. 2004; Yang et al. 2017), enabling investigations into their frequency of use. However, a careful review of prior studies reveals concerns in the application of the SINS II scale.

First, many studies utilized only a subset of the 25 tactics. For instance, Zarkada-Fraser and Fraser (2001) identified only nine tactics as relevant to international business negotiations based on expert panel reviews. Similarly, Volkema et al. (2004) and Fleck et al. (2013) found only four and six tactics, respectively, to be applicable in negotiation simulations. Rivers and Volkema (2013) later proposed an expanded inventory of 75 negotiation tactics, identifying 24 as relevant, but their adjusted scale has not been widely adopted. Additionally, some tactics were excluded by researchers (e.g., Erkuş and Banai 2011; Hershfield et al. 2012) because they were deemed more appropriate than ethically ambiguous. These deviating applications lead to the first research hypothesis:

H1: All items of the SINS II scale are relevant in negotiations, assuming that at least 25% of participants have encountered or applied each item.

Second, the scale has increasingly focused on measuring tactics' perceived appropriateness rather than their frequency of use. While early versions of the SINS II scale assessed both appropriateness and likelihood of application (Lewicki and Stark 1996), later studies abandoned the frequency measure due to high correlations between the two variables (Lewicki and Robinson 1998; Volkema and Fleury 2002). However, this shift and the requirement of respondents to assess the appropriateness for each item, regardless of their familiarity with the item, increases the risk of collecting irrelevant data on tactics with which participants lack practical experience. Consequently, it appears likely that individuals with differing familiarity may rate a tactic's appropriateness based on societal norms or influenced by social desirability rather than based on their own negotiation experience. Therefore, not controlling participants' actual familiarity may jeopardize the content validity of the responses. However, as this high correlation between appropriateness and frequency variables has been tested during the early stages of the scale's development and before its current set of 25 tactics, it appears important to test this possible limitation for the current items of the SINS II scale. This concern leads to the second hypothesis:

H2: Encountered and self-applied frequency of a SINS II scale item significantly correlate with respondents score its appropriateness.

Finally, social desirability bias may influence participants' responses, leading them to underreport their use of EANT while overestimating the appropriateness of commonly applied tactics (King and Bruner 2000; Holtgraves 2004). To address this concern, the third hypothesis is proposed:

H3: Participants' self-reported frequency of applying SINS II scale items differs significantly from the frequency with which they encountered these tactics, depending on the tactics' perceived appropriateness.

The theoretical basis for our further hypotheses is rooted in several interrelated perspectives. Normative ethical theories (e.g., Rest, 1986) provide benchmarks for distinguishing acceptable from unethical practices, while social exchange theory (Blau, 1964;

Homans 1958; Thibaut and Kelley 1959) and behavioral decision-making frameworks (Kahneman and Tversky 1979; Simon 1955) explain how self-interest, reciprocity, and cognitive biases can lead negotiators to adopt tactics that may deviate from established ethical norms. These theories suggest that individuals' perceptions and evaluations of negotiation behavior are influenced not only by inherent ethical standards but also by social context and personal experience. This integrated theoretical perspective underpins the hypotheses that socio-demographic factors (e.g., age, gender, cultural affiliation, and negotiation experience) are linked to the frequency and perception of ethically ambiguous negotiation tactics.

Regarding the current state of research, numerous empirical studies have established that independent variables significantly influence the frequency and perceived appropriateness of EANT. One prominent area of inquiry is the impact of culture on the application of EANT. Cross-cultural comparisons, such as those between the United States and Saudi Arabia (Al-Khatib et al. 2008), Canada and China (Ma 2010; Liu et al. 2019), and the United States and China (Ma et al. 2013), have identified notable differences in how EANT are rated in terms of appropriateness. However, a comparison of less-explored cultures such as Indians and Germans is warranted. Based on Hofstede's (1980) cultural dimensions theory, Indian participants are hypothesized to employ EANT more frequently and perceive them as less inappropriate than German participants. India's high Power Distance (PDI = 77) and moderate Masculinity (MAS = 56) suggest a hierarchical, competitive society where manipulative tactics may be more readily justified. Intermediate Individualism (IDV = 48) reflects a balance between group loyalty and personal gain, further rationalizing the use of such tactics. Low Uncertainty Avoidance (UAI = 40) indicates a higher tolerance for ambiguity and risk (Hofstede Insights 2021).

H4a: Participants indicating Indian cultural belonging apply EANT more frequently than participants indicating German cultural belonging.
H4b: Participants indicating Indian cultural belonging rate EANT as more appropriate than participants indicating German cultural belonging.

Gender has also been examined as an influencing factor on EANT perceptions. Studies by Kray and Haselhuhn (2012) and Tasa and Bell (2017) revealed that women rate all EANT as less appropriate than men, reflecting stricter ethical standards. These tendencies suggest that women may apply fewer EANT in negotiations, due to social and psychological differences that emphasize ethical considerations and relational outcomes (Kray and Thompson 2005; Bowles et al. 2005).

H5a: Women apply EANT less frequently than men.
H5b: Women rate EANT as less appropriate than men.

The relationship between negotiation experience and EANT usage has not been explicitly studied. However, broader research on EANT indicates that as age increases, the frequency and perceived appropriateness of EANT usage, tend to decrease. This may be attributed to life experience fostering emotional regulation and ethical decision-making, reducing impulsive behavior (Carstensen et al. 2003). Experienced negotiators are also more likely to understand the long-term consequences of EANT, opting for rational strategies that sustain relationships and achieve mutually beneficial outcomes (Lewicki et al. 2021).

H6a: Increasing age decreases the frequency of EANT applications.
H6b: Increasing age decreases the appropriateness of EANT.
H7a: Negotiation experience decreases the frequency of EANT application.
H7b: Negotiation experience decreases the appropriateness of EANT.

Negotiation training as an influencing factor has received limited attention in research. However, training often emphasizes ethical practices, strategic planning, and maintaining integrity, equipping individuals with the tools to achieve their goals ethically (Lewicki et al. 2021). Consequently, it is hypothesized that increased negotiation training reduces the frequency and perceived appropriateness of EANT application.

H8a: Increased negotiation training decreases the frequency of EANT application.
H8b: Increased negotiation training decreases the appropriateness of EANT.

The influence of industry on EANT application has also been highlighted in research. For example, Cohn et al. (2014) argue that the "prevailing business culture in the banking industry weakens and undermines the honesty norm" in business practices. If this claim holds, it would imply higher frequency and perceived appropriateness among members of the finance, investment, and insurance industries.

H9a: Members of the finance, investment, and insurance industries apply EANT more frequently than members of other industries.
H9b: Members of the finance, investment, and insurance industries rate EANT as more appropriate than members of other industries.

3 Research Design

3.1 Survey Set-up

To test the proposed research hypotheses, a survey-based quantitative empirical research design was adopted. The survey was developed using Qualtrics as the survey management platform and administered in both English and German. Translation was conducted by three independent researchers, with discrepancies discussed and resolved collaboratively. Subsequently, the German version was back-translated into English by two additional researchers, with further discussions to resolve any inconsistencies and align the translation with the original questionnaire to minimize errors.

Consistent with prior studies, the survey collected basic socio-demographic data, including age, gender, and national identity (as a proxy for culture). To ensure anonymity, age was categorized into 10-year intervals ranging from 20 to 69+. For the first time in negotiation ethics research, additional variables were included: industry affiliation, the number of negotiation training courses attended, and self-reported negotiation experience, measured using a five-point Likert scale.

Following the socio-demographic section, participants were presented with up to three questions for each of the 25 tactics in the SINS II scale. The tactics were randomized to reduce order bias. To assess the relevance of the SINS II scale items, as posited in H1, participants were asked how often they had applied a tactic or experienced it being applied by a counterparty or third party. This approach captured both perspectives in negotiations and enabled verification that, in a sufficiently large sample, the frequency of tactics applied by both sides would converge, assuming a normal distribution of ethical behavior among negotiators. For each SINS II scale item, participants answered the following questions: (1) "How frequently have you seen your counterpart(s) use this tactic in real-life negotiations?" and (2) "How frequently have you applied this tactic in real-life negotiations?" Responses were rated on a five-point Likert scale: 1 = Never, 2 = Rarely, 3 = Sometimes, 4 = Often, and 5 = Always. If participants answered "Never" to both questions, the survey automatically advanced to the next tactic, assuming their lack of experience with the tactic precluded informed evaluation. This adaptive approach was novel compared to prior studies utilizing the SINS II scale, which did not account for participants' familiarity with specific tactics.

For tactics that met the experience threshold, participants assessed their perception of the tactic's appropriateness: (3) "How do you regard the use of this tactic in a negotiation in which something important to you and your business is at stake?" Responses were rated on a five-point Likert scale: 1 = Very inappropriate, 3 = Neither appropriate nor inappropriate, and 5 = Very appropriate.

3.2 Sampling

The study employed two distinct sample groups to ensure a diverse and representative dataset. The first sample consisted of alumni and postgraduate students from a European business school, recruited via LinkedIn. This approach specifically targeted experienced negotiation practitioners, given that business school affiliates are more likely to engage frequently in negotiations. LinkedIn's search function was used to identify individuals affiliated with the school (using the "school" filter) and referencing negotiation-related terms in their profiles. Keywords included "negotiation," "negotiations," "negotiator," or "negotiating," along with translations in German, French, Spanish, and Portuguese (e.g., "Verhandlung," "négociation," "negociación," "negociação"). Attempts to include additional languages (e.g., Russian, Hindi, Mandarin) yielded no results. This process identified 1,381 potential participants, who were invited via personalized connection requests. Of these, 892 accepted and received a neutral description of the study, along with a survey link hosted on Qualtrics. The study was described as examining the frequency and appropriateness of negotiation tactics in business negotiations, without explicit reference to ethical considerations. Up to two reminders (spaced one week apart) were sent. Ultimately, 507 individuals engaged with the survey, 74 of whom later declined participation, yielding 433 responses and a 31% participation rate.

To address potential bias associated with business school affiliation and ensure greater cultural and industry diversity, a second sample was collected through public LinkedIn posts from the three authors, inviting voluntary participation. Respondents in this group were neither alumni nor current students of the aforementioned business school. This method provided access to a broader professional audience, capturing perspectives from different cultural backgrounds and industries. The second sampling method produced 67 completed surveys. Statistical tests revealed no significant differences (at the 95% confidence level) between the two groups in the means of EANT frequency or appropriateness for any variables studied. Consequently, the data from both samples were merged for analysis, resulting in a combined dataset of 500 completed surveys.

To enhance gender balance and ensure representation across negotiation experience levels, demographic quotas were monitored throughout the recruitment process. While the initial sample skewed toward male respondents (reflecting the gender distribution often found in business negotiation roles), additional outreach efforts were made to improve female participation. Despite these efforts, gender imbalance remains a limitation, with male respondents comprising 78% of the final dataset. However, negotiation experience levels were well-distributed, ensuring insights from both novice and highly experienced negotiators.

Participants received no monetary compensation for survey completion, apart from a promised digital copy of the study's findings. To ensure data quality, participation times were examined to identify and exclude implausibly rapid responses. A reasonable completion time was estimated at 60 s for socio-demographic questions and 30 s per SINS II scale item. Five responses failed this criterion and were excluded. The final sample comprised 495 completed surveys, encompassing 12,375 tactic-based data points. Table 1 presents the socio-demographic characteristics of the collected samples.

Table 1. Socio-demographic characteristics of the collected samples.

Age	#	Share	Industry	#	Share
20-29	87	18%	Manufacturing	92	19%
30-39	192	39%	Finance, investment or insurance	71	14%
40-49	156	32%	Information or communication	69	14%
50-59	50	10%	Trade (retail)	23	5%
>=60	10	2%	Others	240	48%
Gender			**Negotiation Trainings**		
Male	385	78%	None	63	13%
Female	110	22%	1	120	24%
			2-4	218	44%
Nationality			5 or more	94	19%
Germany	178	36%			
India	51	10%	**Negotiation Experience**		
U.S.A.	18	4%	1 - None	7	1%
Poland	17	3%	2 - A little	67	14%
Russian Federation	15	3%	3 - A moderate amount	188	38%
Peru	13	3%	4 - A lot	144	29%
57 Other	203	41%	5 - An extensive amount	89	18%

4 Results

4.1 Reliability of the SINS II Scale

To evaluate the relevance of the SINS II scale items and its statistical reliability for measuring EANT, three key aspects were analyzed: internal consistency, construct validity, and content validity. Missing values in Question 3 were adjusted using the Expectation

Maximization algorithm in SPSS. Internal consistency was assessed by Cronbach's alpha coefficient for Questions 1, 2, and 3, yielding high values exceeding the recommended threshold of 0.7: Question 1 = 0.926, Question 2 = 0.892, and Question 3 = 0.855.

To determine construct validity, a factor analysis was performed on adjusted responses to Question 3. The results recommended a seven-factor solution, consistent with the original factor analysis by Lewicki and Robinson (1998) and Fulmer et al. (2009), with all but one tactic aligning with the original factors. This supports the scale's internal consistency and construct validity.

Content validity was examined using responses from Questions 1 and 2. Crosstabulations of SINS II scale items with Likert-scale ratings were converted into percentage-based values for easier interpretation (e.g., 1 – Never = 0%, 2 – Rarely = 25%, 3 – Sometimes = 50%, 4 – Often = 75%, 5 – Always = 100%).

Table 2. Tables of encountered and self-applied frequencies sorted by appropriateness score.

Tactic	Encountered frequency	Sig. T-test	Self-applied frequency	Sig. T-test	Difference	Appropriateness
24. False Guarantees	30,2%	<0.001	13,4%	<0.001	16,7%	1,86
15. Make Threats	23,1%	0,054	9,1%	<0.001	14,0%	1,95
1. Empty Promise	37,0%	<0.001	14,9%	<0.001	22,1%	1,96
5. Exchange Opponent	14,2%	<0.001	7,2%	<0.001	7,0%	2,08
14. Fake Consessions	38,1%	<0.001	18,1%	<0.001	20,1%	2,14
19. Undermine Network	27,5%	0,019	14,9%	<0.001	12,6%	2,22
25. Recruit Opponents	19,7%	<0.001	8,3%	<0.001	11,4%	2,26
3. Misinformation	42,0%	<0.001	24,2%	0,255	17,8%	2,31
21. Bribing Opponent	32,2%	<0.001	16,2%	<0.001	16,0%	2,34
8. Bribe Network	17,0%	<0.001	8,4%	<0.001	8,6%	2,34
18. Misrepresent Principal	34,8%	<0.001	19,6%	<0.001	15,2%	2,42
16. Denying Validity	43,4%	<0.001	26,7%	0,069	16,8%	2,42
6. Misinform Principal	29,3%	<0.001	18,3%	<0.001	11,0%	2,42
11. Fake Disgust	27,8%	0,010	16,8%	<0.001	11,1%	2,51
22. Fake Fury	32,5%	<0.001	18,0%	<0.001	14,5%	2,63
9. Fake Fear	24,6%	0,374	14,2%	<0.001	10,4%	2,64
20. Fake Melancholy	24,7%	0,414	15,7%	<0.001	9,1%	2,70
4. Strategic Anger	38,0%	<0.001	26,1%	0,164	11,9%	2,73
17. Fake Caring	48,1%	<0.001	34,0%	<0.001	14,1%	2,95
23. Intimidating Demands	44,7%	<0.001	32,6%	<0.001	12,1%	3,02
7. Fake Empathy	49,3%	<0.001	41,5%	<0.001	7,9%	3,13
13. Fake Disappointment	48,8%	<0.001	38,6%	<0.001	10,2%	3,19
2. Fake Friendship	53,7%	<0.001	45,9%	<0.001	7,8%	3,43
12. Time Pressure	50,7%	<0.001	47,0%	<0.001	3,7%	3,62
10. Over Anchoring	66,0%	<0.001	59,3%	<0.001	6,7%	3,87

A relevance threshold of 25% was applied, indicating at least rare usage or encounter of the tactic. Results are shown in Table 2 and ranged from 7% to 59% for self-applied frequency and 14% to 66% for encountered frequency. Notably, only 7 of 25 tactics surpassed the self-applied frequency threshold, whereas 19 exceeded the encountered threshold. Six tactics failed to meet the threshold for either question. This variability highlights significant differences in the relevance of the SINS II scale items, raising concerns about content validity. Therefore, H1 is rejected. Despite this limitation, the SINS II scale's internal consistency and construct validity support its continued use in empirical research.

4.2 Subjects' Familiarity with SINS II Scale Items

To examine whether differences in familiarity with SINS II scale items affect overall ratings of their appropriateness, we replicated the analyses performed by Lewicki and Robinson (1998) and Volkema and Fleury (2002) by calculating Spearman's rank correlation coefficient. While those studies used earlier versions of the scale, our study is the first to test the 25-item version that has been in use since 2009. In our sample, the correlation coefficient between encountered frequency and perceived appropriateness across individual tactics was 0.37 ($p < 0.001$), with only the "Exchange Opponent" item showing no significant correlation. For applied frequency and perceived appropriateness, the correlation was even stronger at 0.69 ($p < 0.001$), with all individual tactics demonstrating significant correlations. Therefore, H2 cannot be rejected.

Figure 1 displays these results by mapping the encountered and applied frequencies according to the perceived appropriateness score of the SINS II scale items.

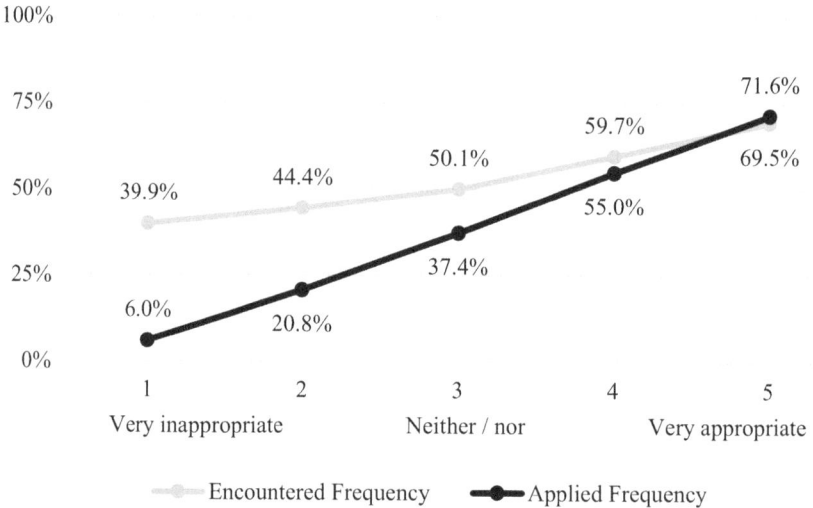

Fig. 1. Mean encountered and applied frequency per group of perceived appropriateness.

The figure clearly illustrates that tactics rated as inappropriate tend to have low encountered frequency and even lower self-applied frequency. If a participant has never applied or encountered a particular tactic, their assessment of its appropriateness—when forced to respond by the survey—may merely reflect prevailing social norms rather than actual negotiation experiences. This calls into question the content validity of such items. Overall, these findings imply that while all tactics are present in negotiations, negotiators' experiences vary considerably and should be taken into account when designing survey instruments.

4.3 Social Desirability Bias

A notable pattern emerged from Table 2: the encountered frequency of all 25 SINS II scale items consistently exceeded the self-applied frequency. This suggests that respondents claimed to act more ethically than their negotiation counterparts. A similar pattern is observed in Fig. 1; for every appropriateness rating except the highest (5 = very appropriate), the encountered frequency was higher than the self-admitted frequency. To investigate this discrepancy, we conducted statistical tests—including the Mann-Whitney U test for gender and the Kruskal-Wallis test for age, training, experience, and appropriateness. Among these factors, only appropriateness showed significant differences. This trend indicates either an overestimation of a counterpart's application of EANT or an underreporting of one's own use due to social desirability bias.

To further explore the possibility of social desirability bias, we created a new variable defined as the difference between encountered and self-applied frequency and mapped this variable against the appropriateness ratings for each data point. As shown in Fig. 2, when the encountered frequency exceeded the self-applied frequency, the appropriateness rating was consistently below the average value of three—indicating that these tactics were regarded as inappropriate. Conversely, when the frequencies were equal or the self-applied frequency was higher, the appropriateness rating was above average, with higher ratings corresponding to a greater surplus in self-applied frequency.

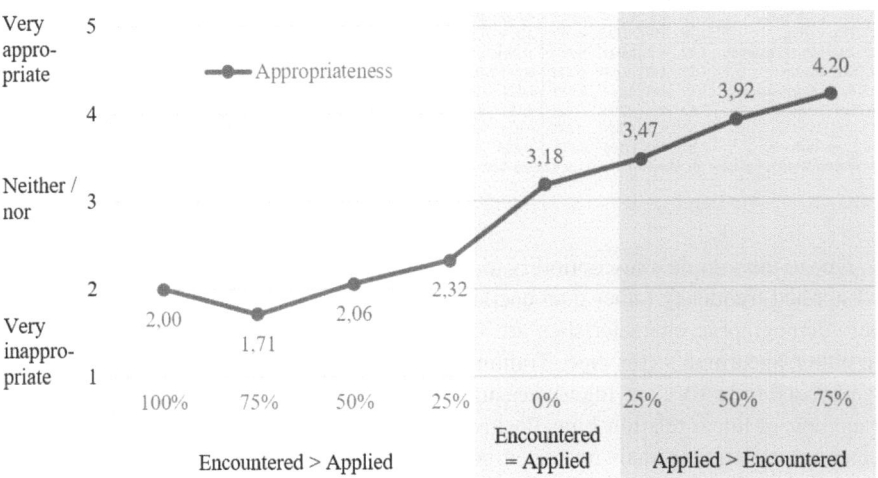

Fig. 2. Appropriateness per group of difference between encountered and applied frequency.

Combined with the findings from Fig. 1, these results support H3 and illustrate that participants may be caught in a dilemma when rating ethical behavior while simultaneously reporting their own involvement in such tactics.

4.4 Socio-demographic Variables

The final analysis examined the relationships between socio-demographic variables, EANT frequency, and perceived appropriateness. As summarized in Table 3, the results are sorted in ascending order based on the mean appropriateness scores from Question 3. The table also indicates the number of responses per tactic, noting that answers were only required if responses to Questions 1 and 2 were not "never." For each independent variable, two columns are presented: one for self-applied frequency (AFQ) and one for appropriateness (APR).

Table 3. Correlation coefficients of socio-demographic variables with the applied frequency and appropriateness per SINS II scale item sorted by appropriateness.

Tactic	Value	N	Gender		Age		Training		Experience		Industry		N	Culture	
			AFQ°	APR	AFQ°	APR	AFQ°	APR	AFQ°	APR	AFQ°	APR		AFQ°	APR
24. False Guarantees	1,86	348	0,031	-0,008	-0,048	-0,054	0,109	-0,012	**0,023**[+]	-0,050	0,015	0,010	159	**0,206***	**-0,149**[+]
15. Make Threats	1,95	278	0,029	0,028	-0,003	**-0,085**[+]	0,016	0,025	0,012	-0,046	0,040	0,040	114	0,031	**-0,188**[+]
1. Empty Promise	1,96	369	0,025	-0,006	-0,055	-0,104	0,083	0,049	-0,048	-0,066	-0,024	0,016	161	**0,213**[+]	0,083
5. Exchange Opponent	2,08	214	**0,109***	0,101	**0,121***	0,048	0,062	0,074	**0,131**[+]	0,111	0,057	0,055	95	**0,237***	0,018
14. Fake Concessions	2,14	387	0,052	-0,032	**-0,118**[+]	-0,107	0,031	0,042	-0,047	-0,024	-0,011	0,035	172	**0,303***	0,031
19. Undermine Network	2,22	317	0,021	0,023	0,018	0,047	0,081	0,007	0,086	0,095	0,036	0,043	134	**0,155***	-0,010
25. Recruit Opponents	2,26	230	0,004	0,019	-0,013	-0,055	0,058	0,030	**-0,057**[+]	0,053	0,013	0,011	102	0,015	**-0,195***
3. Misinformation	2,31	404	-0,006	0,007	-0,010	-0,002	0,084	0,045	0,043	-0,021	0,050	0,013	176	**0,214***	0,047
21. Bribing Opponent	2,34	334	0,006	0,024	-0,026	**-0,127**[+]	0,074	0,004	0,022	-0,031	-0,054	-0,004	135	0,072	-0,035
8. Bribe Network	2,34	202	0,018	-0,015	-0,029	-0,029	-0,005	0,004	-0,076	-0,037	-0,016	-0,058	83	**0,162**[+]	0,038
18. Misrepresent Principal	2,42	355	-0,033	-0,034	-0,012	-0,030	0,071	0,043	0,033	0,008	-0,019	0,029	157	**0,198***	-0,021
16. Denying Validity	2,42	420	0,036	0,014	-0,010	0,043	0,076	0,021	0,103	0,067	-0,036	0,014	189	**0,160***	-0,021
6. Misinform Principal	2,42	324	0,059	-0,026	-0,017	-0,090	-0,011	-0,012	0,001	-0,086	0,012	0,029	144	**0,264***	0,029
11. Fake Disgust	2,51	307	0,004	0,057	0,089	0,037	**0,113***	0,048	0,086	0,056	-0,048	0,005	126	**0,218***	-0,030
22. Fake Fury	2,63	358	0,070	0,023	0,036	0,049	0,057	-0,038	0,060	0,018	0,022	0,010	166	0,107	**-0,162**[+]
9. Fake Fear	2,64	306	-0,008	0,001	-0,001	-0,008	0,047	-0,007	-0,013	-0,091	-0,011	**-0,134**[+]	130	**0,126**[+]	-0,151
20. Fake Melancholy	2,70	295	-0,001	0,023	**0,026**[+]	0,008	0,049	-0,097	**0,012***	-0,024	0,054	0,109	115	**0,375***	0,131
4. Strategic Anger	2,73	400	**0,120**[+]	**0,097**[+]	0,027	0,029	**0,115**[+]	0,023	**0,147***	0,086	-0,022	0,058	180	-0,072	**-0,419***
17. Fake Caring	2,95	419	-0,085	**-0,094**[+]	**-0,122**[+]	**-0,144**[+]	0,039	0,057	0,002	0,027	-0,062	**-0,128***	184	**0,157**[+]	-0,068
23. Intimidating Demands	3,02	411	**0,094**[+]	0,084	-0,064	**-0,165**[+]	0,073	0,020	-0,032	**-0,151**[+]	-0,033	-0,094	184	0,015	**-0,186***
7. Fake Empathy	3,13	443	-0,049	**-0,088**[+]	**-0,147***	**-0,128**[+]	0,057	0,052	-0,002	-0,063	0,009	0,021	199	**0,241***	-0,128
13. Fake Disappointment	3,19	444	**0,114**[+]	**0,087**[+]	**0,079**[+]	-0,016	**0,154***	0,063	**0,160***	**0,091**[+]	-0,001	-0,034	203	**0,277***	**0,170**[+]
2. Fake Friendship	3,43	454	0,053	0,026	**-0,197***	**-0,189***	0,046	0,011	0,012	**-0,008***	-0,023	-0,039	204	**0,189***	0,139
12. Time Pressure	3,62	445	0,076	0,048	0,034	-0,027	0,051	0,019	**0,089**[+]	0,100	0,027	0,066	201	0,134	0,027
10. Over Anchoring	3,87	473	**0,106**[+]	**0,129**[+]	-0,090	-0,052	**0,113**[+]	0,026	0,036	0,006	-0,029	-0,031	221	**0,313***	-0,017

° Constant N=495; + p-value for Mann-Whitney U or Kruskal-Wallis test <0,05; * p-value for Mann-Whitney U or Kruskal-Wallis test <0,01

Due to the sample's susceptibility to social desirability bias, the analysis focuses on self-applied frequency rather than encountered frequency, since only the respondents' socio-demographic characteristics are known. The displayed correlation coefficients are either Spearman's (for Age, Training, and Experience) or Pearson's (for Culture, Gender, and Industry), serving as measures of the direction and strength of the assumed monotonic or linear relationships. Positive correlations are highlighted in grey, whereas negative correlations remain unshaded. Additionally, nonparametric tests—the Kruskal-Wallis test and the Mann-Whitney U test—were conducted for each relationship to compare the distributions of the independent groups' means. When the p-value of these tests falls below 0.05 or 0.01, the corresponding correlation coefficient is highlighted in bold and marked with a symbol. Overall, the results can be summarized as follows:

- **Cultural Belonging**: Indian participants applied EANT more frequently than German participants (H4a), supported by 18 of 25 tactics showing statistical significance. However, H4b (Indian participants rate EANT as more appropriate) was not supported, with only one tactic showing significance.
- **Gender**: H5a (women apply EANT less frequently than men) was supported for five tactics but inconclusive for the rest. H5b (women rate EANT as less appropriate than

men) was mostly inconclusive, with three tactics showing significance and contradicting previous studies suggesting that women rate the appropriateness of all EANT lower than men (Kray and Haselhuhn 2012; Tasa and Bell 2017).
- **Age**: H6a (increased age decreases EANT frequency) was supported for four tactics, while most results were inconclusive. H6b (increased age decreases EANT appropriateness) was supported for six tactics.
- **Negotiation Experience**: H7a (experience decreases EANT frequency) was supported for one tactic, while H7b (experience decreases appropriateness) was supported for two tactics.
- **Negotiation Training**: H8a (training decreases EANT frequency) was unsupported, with four tactics showing significant rejection, which could hint that they are even promoted as tools with less ethical concerns in negotiation training. H8b (training decreases appropriateness) was inconclusive.
- **Industry**: H9a (finance industry members apply EANT more frequently) and H9b (finance industry members rate EANT as more appropriate) were inconclusive.

Notably, our non-parametric analysis uncovered a fourth limitation of the SINS II scale. Aggregating data by the seven tactic groups—or by using the total questionnaire scores, as previous studies often have done—risks obscuring the nuanced interactions between socio-demographic factors and individual tactics. This limitation highlights the need for a more granular analytical approach when applying the SINS II scale in negotiation ethics research.

5 Discussion

Our analysis of the SINS II scale demonstrated high internal consistency and construct validity, affirming its utility as a tool for measuring EANT. However, the low frequency values for certain scale items raise concerns about content validity. Negotiators may interpret tactics differently due to personal biases, experiences, and situational factors, highlighting the need for a more nuanced approach to assessing ethical ambiguity in negotiations. Incorporating measures of participants' familiarity or interaction frequency with specific EANT could enhance the scale's applicability. Without such measures, participants may be forced to express opinions about unfamiliar tactics, potentially skewing results and masking the true ethical implications of these behaviors.

The examination of socio-demographic factors—including gender, age, negotiation experience, training, industry, and culture—produced mixed results regarding their influence on EANT frequency and appropriateness. Contrary to earlier studies, gender did not significantly affect EANT frequency or appropriateness, suggesting a more complex relationship between gender and negotiation behavior than previously assumed. Age, however, emerged as a significant factor, with older negotiators reporting lower frequency and appropriateness of EANT, possibly reflecting generational differences in negotiation styles.

Negotiation experience and training were associated with higher EANT frequency, though they did not significantly impact perceptions of appropriateness. This finding suggests that familiarity with negotiation contexts may normalize ethically ambiguous tactics over time. Such results are particularly relevant for negotiation educators, as they underscore the paradox that increased awareness of unethical tactics might inadvertently encourage their application. Instructors aiming to foster ethical negotiation practices should consider this potential normalization effect when designing curricula.

Cultural differences significantly influenced EANT frequency and appropriateness, reinforcing the contextual nature of negotiation behavior. These results emphasize the importance of situational factors in ethical decision-making and warrant further exploration in diverse cultural settings. Expanding research to include underrepresented regions such as Africa, Central Asia, and East Asia could provide a more comprehensive understanding of global negotiation ethics. Some first publications, such as Ma et al. (2023) started the research in the Asian domain by applying the SINS II scale for the comparison of Indian and Pakistani negotiation styles.

The analysis identified a potential social desirability bias in self-reported EANT frequency, as participants tended to report lower application rates compared to their perceived encounter rates—a discrepancy that likely reflects a reluctance to admit engagement in ethically questionable behaviors. To mitigate this bias, future implementations of the SINS II scale should adopt neutral question framing and incorporate a third-party negotiator's perspective. This approach minimizes respondents' tendency to personally identify with the ethical dilemma, which is further supported by the observed lower correlation between encountered frequency and appropriateness. Such a carefully considered research design not only limits social desirability bias but also enhances the accuracy of self-reported data, suggesting broader applicability in future research.

In addition to these empirical findings, the study offers several theoretical contributions that extend existing models of negotiation ethics. First, the findings challenge conventional assumptions that increased negotiation experience and training uniformly lead to more ethical behavior. Instead, our results suggest that repeated exposure to negotiation contexts may normalize the use of ethically ambiguous tactics, calling for a re-evaluation of the relationship between experience and ethical decision-making. Second, by revealing the complex influence of socio-demographic factors, our study supports an integrative theoretical framework that accounts for both individual and contextual determinants of ethical behavior in negotiations. This nuanced perspective underscores the need for negotiation ethics theories to incorporate cultural, generational, and situational variability rather than relying solely on aggregated measures. Finally, the observed discrepancies attributable to social desirability bias invite further theoretical inquiry into the measurement challenges inherent in studying unethical behavior. Future research that employs mixed-method approaches could refine these theoretical models by bridging the gap between self-reported data and observed behavior, thereby offering a more holistic understanding of ethical ambiguity in negotiation settings.

6 Future Research Directions and Limitations

This study opens several avenues for future research on negotiation behavior and ethical decision-making. Longitudinal studies examining how negotiation behavior evolves over time and its interaction with socio-demographic factors could offer valuable insights into the stability and variability of ethical decision-making. Additionally, while efforts were made to recruit a diverse participant pool, the sample was predominantly European, with a strong focus on German and Indian participants. This limits the generalizability of the findings to other cultural contexts, particularly in regions where negotiation norms and ethical considerations may differ significantly. Future research should expand sampling to underrepresented regions, such as Africa, Central Asia, and East Asia, to capture a broader range of negotiation practices and ethical perspectives.

Another key limitation of this study lies in its reliance on survey-based methods, which may amplify social desirability bias and obscure participants' subconscious actions during negotiations. Future research employing observation-based methods or experimental setups could provide deeper insights into the implicit and explicit dynamics of ethical decision-making in negotiations, reducing reliance on self-reported data and uncovering behavioral patterns that participants may not consciously recognize or admit.

By addressing these limitations, future studies can contribute to a more comprehensive and globally applicable understanding of ethical decision-making in negotiations.

Disclosure of Interests. The authors have no competing interests to declare that are relevant to the content of this article.

References

Al-Khatib, J.A., Malshe, A., Abdul Kader, M.: Perception of unethical negotiation tactics: a comparative study of US and Saudi managers. Int. Bus. Rev. **17**(1), 78–102 (2008)

Aquino, K., Becker, T.E.: Lying in negotiations: how individual and situational factors influence the use of neutralization strategies. J. Organ. Behav. Int. J. Ind. Occup. Organ. Psychol. Behav. **26**(6), 661–679 (2005)

Banas, J.T., Parks, J.M.: Lambs among lions? The impact of ethical ideology on negotiation behaviors and outcomes. Int. Negot. **7**(2), 235–260 (2002)

Blau, P.M.: Exchange and Power in Social Life. Wiley, New York, NY (1964)

Boles, T.L., Croson, R.T., Murnighan, J.K.: Deception and retribution in repeated ultimatum bargaining. Organ. Behav. Hum. Decis. Process. **83**(2), 235–259 (2000)

Bowles, H.R., Babcock, L., McGinn, K.L.: Constraints and triggers: situational mechanics of gender in negotiation. J. Pers. Soc. Psychol. **89**(6), 951–965 (2005)

Carson, T.L., Wokutch, R.E., Murrmann, K.F.: Bluffing in labor negotiations: legal and ethical issues. J. Bus. Ethics **1**(1), 13–22 (1982)

Carstensen, L.L., Fung, H.H., Charles, S.T.: Socioemotional selectivity theory and the regulation of emotion in the second half of life. Motiv. Emot. **27**(2), 103–123 (2003)

Cohn, A., Fehr, E., Maréchal, M.A.: Business culture and dishonesty in the banking industry. Nature **516**(7529), 86–89 (2014)

Dees, J.G., Cramton, P.C.: Shrewd bargaining on the moral frontier: toward a theory of morality in practice. Bus. Ethics Q. **1**(2), 135–167 (1991)

Erkuş, A., Banai, M.: Attitudes towards questionable negotiation tactics in Turkey. Int. J. Confl. Manag. **22**(3), 239–263 (2011)

Fisher, R., Ury, W., Patton, B.: Getting to Yes: Negotiating Agreement Without Giving In, 2nd edn. Penguin Books, New York (1991)

Fleck, D., Volkema, R., Levy, B., Pereira, S., Vaccari, L.: Truth or consequences: the effects of competitive-unethical tactics on negotiation process and outcomes. Int. J. Confl. Manag. **24**(4), 328–351 (2013)

Friedman, R.A., Shapiro, D.L.: Deception and mutual gains bargaining: are they mutually exclusive? Negot. J. **11**(3), 243–247 (1995)

Fulmer, I., Barry, B., Long, D.: Lying and smiling: informational and emotional deception in negotiation. J. Bus. Ethics **88**(4), 691–709 (2009)

Gino, F., Shu, L.L., Bazerman, M.H.: Nameless + harmless = blameless: when seemingly irrelevant factors influence judgment of (un)ethical behavior. Organ. Behav. Hum. Decis. Process. **111**(2), 93–101 (2010)

Hershfield, H.E., Cohen, T.R., Thompson, L.: Short horizons and tempting situations: lack of continuity to our future selves leads to unethical decision making and behavior. Organ. Behav. Hum. Decis. Process. **117**(2), 298–310 (2012)

Hofstede Insights. https://www.hofstede-insights.com/country-comparison/. Accessed 30 May 2024

Hofstede, G.: Culture's Consequences: International Differences in Work-Related Values. Sage, Beverly Hills (1980)

Homans, G.C.: Social behavior as exchange. Am. J. Sociol. **63**(6), 597–606 (1958)

Holtgraves, T.: Social desirability and self-reports: testing models of socially desirable responding. Pers. Soc. Psychol. Bull. **30**(2), 161–172 (2004)

Kahneman, D., Tversky, A.: Prospect theory: an analysis of decision under risk. Econometrica **47**(2), 263–291 (1979)

King, M.F., Bruner, G.C.: Social desirability bias: a neglected aspect of validity testing. Psychol. Mark. **17**(2), 79–103 (2000)

Kray, L.J., Haselhuhn, M.P.: Male pragmatism in negotiators' ethical reasoning. J. Exp. Soc. Psychol. **48**(5), 1124–1131 (2012)

Kray, L.J., Thompson, L.: Gender stereotypes and negotiation performance: an examination of theory and research. Res. Organ. Behav. **26**, 103–182 (2005)

Lax, D.A., Sebenius, J.K.: Three ethical issues in negotiation. Negot. J. **2**(4), 363–370 (1986)

Lewicki, R.J., Robinson, R.J.: Ethical and unethical bargaining tactics: an empirical study. J. Bus. Ethics **17**(6), 665–682 (1998)

Lewicki, R.J., Saunders, D.M., Barry, B.: Essentials of Negotiation. 7th edn. McGraw-Hill Education, New York, NY (2021)

Lewicki, R.J., Stark, N.: What is ethically appropriate in negotiations: an empirical examination of bargaining tactics. Soc. Just. Res. **9**(1), 69–95 (1996)

Liu, X., Ma, Z., Liang, D.: Personality effects on the endorsement of ethically questionable negotiation strategies: business ethics in Canada and China. Sustainability **11**, 3097–3115 (2019)

Ma, Z., Li, K., Guo, G., Pathak, J., Song, Y.H.: Ethically questionable negotiation strategies in South Asia: a comparative study of India and Pakistan. Group Decis. Negot. **32**, 1289–1314 (2023)

Ma, Z., Liang, D., Chen, H.: Negotiating with the Chinese: are they more likely to use unethical strategies? Group Decis. Negot. **22**(4), 641–655 (2013)

Ma, Z.: The SINS in business negotiations: explore the cross-cultural differences in business ethics between Canada and China. J. Bus. Ethics **91**(1), 123–135 (2010)

McCornack, S.A., Levine, T.R.: When lies are uncovered: emotional and relational outcomes of discovered deception. Commun. Monogr. **57**(2), 119–138 (1990)

Michelman, J.H.: Deception in commercial negotiation. J. Bus. Ethics **2**(4), 255–262 (1983)

Olekalns, M., Horan, C.J., Smith, P.L.: Maybe it's right, maybe it's wrong: structural and social determinants of deception in negotiation. J. Bus. Ethics **122**(1), 89–102 (2014)

Rest, J.: Moral Development: Advances in Research and Theory. Praeger, New York (1986)

Rivers, C., Volkema, R.J.: East-west differences in 'Tricky' Tactics: a comparison of the tactical preferences of Chinese and Australian negotiators. J. Bus. Ethics **115**(1), 17–31 (2013)

Robinson, R.J., Lewicki, R.J., Donahue, E.M.: Extending and testing a five factor model of ethical and unethical bargaining tactics: Introducing the SINS scale. J. Organ. Behav. **21**(6), 649–664 (2000)

Rottenburger, J.R., Kaufmann, L.: Picking on the new kid: firm newness and deception in buyer-supplier negotiations. J. Purch. Supply Manag. **26**(1), 1–10 (2020)

Schweitzer, M.E., Brodt, S.E., Croson, R.T.: Seeing and believing: visual access and the strategic use of deception. Int. J. Confl. Manag. **13**(3), 258–375 (2002)

Schweitzer, M.E., Croson, R.T.: Curtailing deception: the impact of direct questions on lies and omissions. Int. J. Confl. Manag. **10**(3), 225–248 (1999)

Schweitzer, M.E., DeChurch, L.A., Gibson, D.E.: Conflict frames and the use of deception: are competitive negotiators less ethical? J. Appl. Soc. Psychol. **35**(10), 2123–2149 (2005)

Shell, G.R.: When is it legal to lie in negotiations? MIT Sloan Manag. Rev. **32**(3), 93–101 (1991)

Simon, H.A.: A behavioral model of rational choice. Q. J. Econ. **69**(1), 99–118 (1955)

Tasa, K., Bell, C.: Effects of implicit negotiation beliefs and moral disengagement on negotiator attitudes and deceptive behavior. J. Bus. Ethics **142**(1), 169–183 (2017)

Tenbrunsel, A.E.: Misrepresentation and expectations of misrepresentation in an ethical dilemma: the role of incentives and temptation. Acad. Manag. J. **43**(1), 330–339 (1998)

Thibaut, J., Kelley, H.H.: The Social Psychology of Groups. Wiley, New York (1959)

Volkema, R.J., Fleck, D., Hofmeister-Toth, A.: Ethicality in negotiation: an analysis of attitudes, intentions, and outcomes. Int. Negot. **9**(2), 315–339 (2004)

Volkema, R.J., Fleury, M.T.L.: Alternative negotiating conditions and the choice of negotiation tactics: a cross-cultural comparison. J. Bus. Ethics **36**(4), 381–398 (2002)

Yang, Y.D., Cremer, D., Wang, C.: How ethically would Americans and Chinese negotiate? The effect of intra-cultural versus inter-cultural negotiations. J. Bus. Ethics **145**(3), 659–670 (2017)

Zarkada-Fraser, A., Fraser, C.: Moral decision making in international sales negotiations. J. Bus. Ind. Mark. **16**(4), 274–293 (2001)

Evolution of Society Caused by Collective and Individual Decisions
A ViSE Model Study

Pavel Chebotarev[✉]

Technion–Israel Institute of Technology, 3200003 Haifa, Israel
pavel4e@technion.ac.il, pavel4e@gmail.com

Abstract. Decision-making societies may vary in their level of cooperation and degree of conservatism, both of which influence their overall performance. Moreover, these factors are not fixed—they can change based on the decisions agents in the society make in their interests. But can these changes lead to cyclical patterns in societal evolution? To explore this question, we use the ViSE (Voting in Stochastic Environment) model. In this framework, the level of cooperation can be measured by group size, while the degree of conservatism is determined by the voting threshold. Agents can adopt either individualistic or group-oriented strategies when voting on stochastically generated external proposals. For Gaussian proposal generators, the expected capital gain (ECG)—a measure of agents' performance—can be expressed in standard mathematical functions. Our findings show that in neutral environments, societal evolution with open or democratic groups can follow cyclic patterns. We also find that highly conservative societies or conservative societies with low levels of cooperation can evolve into liberal (less conservative than majoritarian) societies and that mafia groups never let their members go when they want to.

1 Introduction

Settings in which agents vote for projects formulated in terms of their capital gains were analyzed by A.V. Malishevskii in the early 1970s; some of these results were presented in [19] and [2]. In Malishevskii's model, voting is easily manipulated by agenda-setters. Similar results have been obtained in frameworks in which participants have ideal points in a multidimensional program space [18]. Spatial models of voting have been studied extensively in [9]. Problems of the relationship between egoism, altruism, collectivism, and rationality were considered in [13,14,17] and voting as a method for making decisions about redistribution of social benefits by means of taxation and social programs was discussed in [10,12,21,22] among others.

Two important factors that influence the performance of decision-making societies are the level of cooperation and the degree of conservatism. Therefore, agents may want to change them in their own interests. What could be the trajectories of such successive reforms? In this paper, we study this question using the ViSE model. Within its framework, the level of cooperation and the degree of conservatism can be measured by the group size and the voting threshold, respectively. We find that the evolution of a society in terms of these parameters can have cyclical and some other specific patterns.

Recent work related to our study is discussed in [3].

1.1 The ViSE Model

The main assumptions of the ViSE model (see [3,15]) are as follows. A *society* consists of n *agents* (also called *voters*). Each agent is characterized by the current value of their *capital* (a real number; debt, if negative), which can sometimes be interpreted as utility. In each step, some proposal is put to the vote, and the agents vote, guided by their strategies, for or against it. Within the ViSE model, a strategy is understood as an algorithm following which an agent uses the available information to decide whether to vote for or against the proposal under consideration. Each *proposal* is a vector of algebraic capital gains of all agents. Proposals approved using the established voting rule are implemented.

Proposals are generated stochastically: their components are realizations of random variables. We consider the case where the components are independent and identically distributed with a given mean μ. The corresponding scalar random variable ξ is called the *gain generator*. The proposals put to the vote can be called *stochastic environment proposals*. The environment is *favorable* if $\mu > 0$, *neutral* if $\mu = 0$, and *unfavorable* (hostile) if $\mu < 0$.

The dynamics of agents' capital in various environments can be analyzed to compare voting strategies and social decision rules in order to select the optimal ones in terms of maximizing appropriate criteria. Since the process in the presented version of the model is stationary, it can be described by one-step characteristics. An important one is the mathematical expectation of one-step capital gain for an agent with a given strategy after implementing the collective decision regarding the submitted proposal using the established voting rule. This value is called the *expected capital gain* (ECG) *of an agent* (this abbreviation was used in [20,23,25], etc.) and denoted by $\mathbb{E}(d_\xi^{\text{str}})$, where abbreviations of agent strategies can be substituted for str. For the societies considered in this paper, the ECG is determined by Corollary 1 formulated in Sect. 2.

The properties of the environment influence the relationship between the current and future states of society. In this regard, the model is relevant to situations where the issue is one of comparing the *status quo* with reform, rather than choosing among several candidates.

This paper examines the decisions made with two agent strategies. An *individualist* (1-agent) supports a proposal if and only if it increases their own capital. The voting strategy of a *group member* is to support a proposal if and only if this proposal increases the group's total capital. We consider a society in which ℓ participants are individualists and g agents form a group, so that $g + \ell = n$ is the size of the society. We denote this society $I|G$ and call it a *single-group* or *two-component* society. A society is *individualistic* (respectively, *clique-like*) if $\ell = n$ (resp., $g = n$).

The votes are aggregated using the t-majority rules: a proposal is accepted if and only if more than t ($t \in [-1, n]$ is a parameter) participants vote for it. Thus, not only a [simple] *majority* ($t = \lfloor \frac{n}{2} \rfloor$) is considered, but also *qualified majorities* ($t \geq \lfloor \frac{n}{2} \rfloor + 1$) often used to make constitutional or other important decisions, and *initiative minorities* with $t < \lfloor \frac{n}{2} \rfloor$, used for such decisions as including issues on the agenda, forming new parliamentary groups, submitting requests, initiating referendums, etc.

Societies will be called *conservative* (resp., *majoritarian*, *liberal*) if $t \geq \lfloor \frac{n}{2} \rfloor + 1$ (resp., $\lfloor \frac{n}{2} \rfloor \leq t < \lfloor \frac{n}{2} \rfloor + 1$, $t < \lfloor \frac{n}{2} \rfloor$); t measures the degree of conservatism. The parameter g expresses the degree of *cooperation* in society. A relative measure of cooperation is g/n. Thus, we classify societies in terms of conservatism and cooperation.

One of the questions under study is: How does the effectiveness of the t-majority voting and the effectiveness of the individualistic and group strategies depend on the parameters t, n, ℓ, and the favorability of the environment?

For a discussion of the relation of the ViSE model to reality, we refer to [15]. Connections with other models are indicated in [3]. Early results on the ViSE model (studying the ECG in two-component societies with different group types and voting thresholds) were obtained in [5]. For more results on two-component societies, see [4]. The ViSE model implements [6] the well-known "small party bias": in the presence of two large parties or coalitions, neither of which has a majority, a small party forming a majority with any of them gains an advantage, which can be comparable to a dictatorship when large coalitions take opposite positions. The model also reveals one of the stability sources of the bipartisan system with almost equal parties. The ViSE model implements [4,6] the "snowball" scenario of cooperation: since belonging to a group is usually more beneficial than protecting one's own interests individually, participants join the group, and as it grows, group egoism becomes closer to altruism. In [7,16], the pit of losses paradox was discussed and the problem of the optimal voting threshold has been solved. A "responsible elite" can help overcome the pit of losses paradox [23,24]. Another possible solution is based on taxes [1]. In [8,15], it was found that the tail heaviness of the distribution of random variables generating proposals affects the effectiveness of agent's strategies.

1.2 Research Framework

It is not uncommon for people to unite into teams or groups and defend group interests rather than individual ones. In fact, this is the background to almost every historical event. Group interest may differ greatly from individual interest, so the question of whether this strategy is rational is non-trivial. This issue is studied within various models including game-theoretic ones. The ViSE model is useful for such studies, as it provides convenient means for varying essential parameters, such as the favorability of the environment, the structure of the society, the social decision rule, etc. In this paper, we study how the performance of individualistic and group agents depends on the voting threshold and the structure of a two-component society.

Furthermore, the teams/groups can appear, increase, decrease, merge, split, and disappear. Individualists may join groups, while group members may become individualists or form other groups when this is beneficial to them. Finally, changing the voting threshold t can be beneficial for some categories of agents and disadvantageous for others. Decisions on changing t can be made by majority voting or voting with the current threshold. In this paper, we study the consequences and benefits of such transitions. The decisions mentioned determine the dynamics of society.

This dynamics essentially depends on the type of the group. An *open* group accepts all individualists willing to join it and lets its members go whenever they wish. A *democratic* group always allows *leaving* it and allows *entering* it iff it is beneficial to the current members of the group. A *mafia*-group only allows changes to it that are beneficial for its current members. The study found that a mafia group *never* lets go of its members who want to leave.

The purpose of this paper is to explore the possible evolution of such societies. In particular, can the evolution patterns be cyclical? We show that this is the case. For example, a society with a small group and low voting threshold will approve of raising its threshold, then increase the size of the group while gradually lowering the threshold, then support a decrease in the size of the group, which will return the society to its original position.

The study is based on Corollary 1 presented in Sect. 2. We consider societies consisting of $n = 25$ agents. It was found that for larger n, the main regions of the diagrams remain the same, while additional details may appear at the boundaries of the regions. For smaller n, some details disappear. We consider the Gaussian proposal generator as the simplest one. The difference in results for other continuous symmetric distributions is not fundamental; for more information on the dependence on distributions, see [8,15].

2 Expected Capital Gains in a Single-Group Society

Corollary 1 ([4]). *For a society of n agents with a group size g and a voting threshold t, under a Gaussian proposal generator $N^{\mu,\sigma}$, the ECGs of individualists and group members are expressed as follows:*

$$\mathbb{E}(d^{\text{ind}}) = \left(r(P_{t-g-1,\ell-1} - P_{t-g,\ell}) + \mu P_{t-g,\ell}\right) N^{0,1}(\mu_g)$$
$$+ \left(r(P_{t-1,\ell-1} - P_{t,\ell}) + \mu P_{t,\ell}\right) N^{0,1}(-\mu_g);$$

$$\mathbb{E}(d^{\text{gr}}) = (P_{t-g,\ell} - P_{t,\ell})\left(\mu N^{0,1}(\mu_g) + \sigma_g f^{0,1}(\mu_g)\right) + \mu P_{t,\ell},$$

where

$$P_{k,m} = \begin{cases} 1, & k < -1 \\ \sum_{i=k+1}^{m} \binom{m}{i} p^i q^{m-i} = 1 - \text{Bi}(k \mid m, p), & -1 \leq k < m \\ 0, & k \geq m, \end{cases} \quad (1)$$

Bi$(\cdot \,|\, m, p)$ is the binomial CDF with m trials and success probability p, $p = 1 - q$, $q = N^{0,1}\!\left(\frac{\mu}{\sigma}\right)$, $r = \frac{\sigma \tilde{f}}{q}$, $\tilde{f} = f^{0,1}\!\left(\frac{\mu}{\sigma}\right)$, $\mu_g = \frac{\mu}{\sigma_g}$, $\sigma_g = \frac{\sigma}{\sqrt{g}}$, $f^{0,1}(\cdot)$ and $N^{0,1}(\cdot)$ are the standard normal PDF and CDF, respectively.

Remark 1. Using the connection between the binomial and Beta distributions we have in (1): $1 - \text{Bi}(k\,|\,m, p) = \text{B}_{k+1, m-k}(p)$, where $\text{B}_{\alpha, \beta}(\cdot)$ is the CDF of the Beta distribution with shape parameters $\alpha > 0$ and $\beta > 0$.

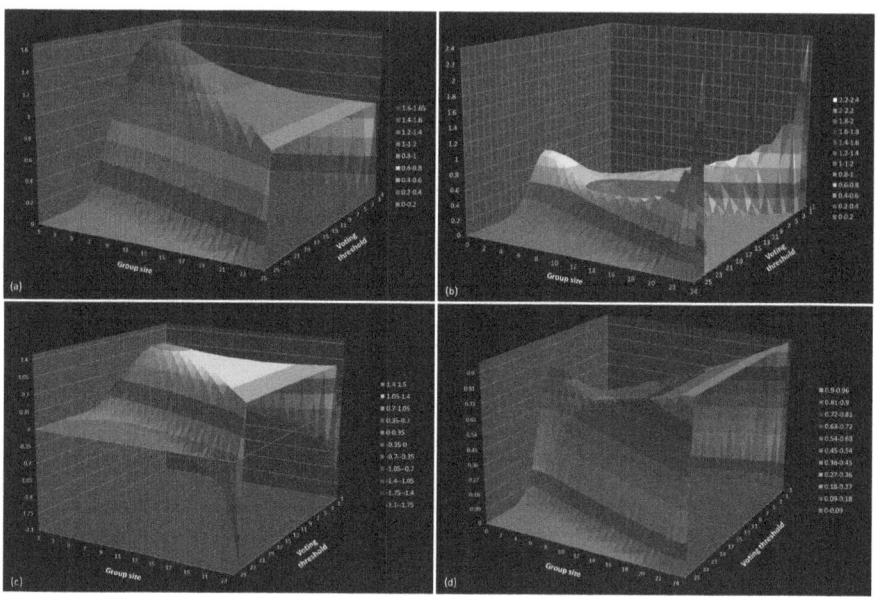

Fig. 1. The ECG of: (a) group members; (b) individualists; (c) difference between them; (d) society's ECG. $I|G$ societies consist of $n = 25$ agents under $N^{0,12}$ generator(Color figure online).

The performance of the two strategies under study is illustrated in Fig. 1. These surfaces have been discussed in [4].

3 Approved Transitions in Two-Component Societies

In this section, we study transformations of society aimed at increasing the welfare of agents. There are two types of changes of interest: structural and procedural. Structural changes of a single-group society are an increase or decrease in the group size. Procedural transformations are changes in the voting threshold. *Elementary transitions* are an increase or decrease in g or t by 1.

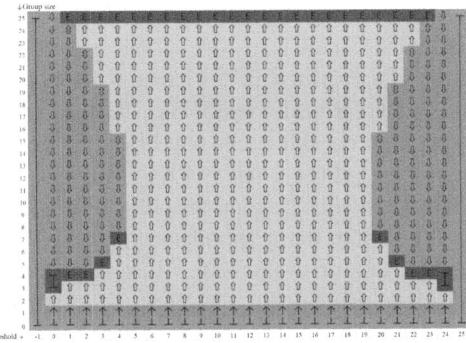

Transitions that change the structure of society				Transitions that change the voting threshold			
Up / Down	Disapproved	Neutral	Approved	Right / Left	Disapproved	Neutral	Approved
Disapproved	⥮	⊥	⇑	Disapproved	⇌	⊢	⇒
Neutral	⊤		↑	Neutral	⊣	—	→
Approved	⇊	↓	⇕	Approved	⇐	←	⇔

Fig. 2. Changes in the degree of cooperation that benefit *defectors* (agents who change their strategy); symbols of transition between societies. (Color figure online)

3.1 Types of Transitions Between Societies

We will analyze diagrams similar to the one shown in Fig. 2. Each cell of a diagram corresponds to a society characterized by coordinates t (voting threshold, on the x-axis) and g (group size, on the y-axis).

The transition symbols are collected in Fig. 2. Each symbol in the left half of the table encodes a *pair* of desirability/undesirability/neutrality indicators associated with transitions in both directions corresponding to a decrease or increase in the group size g by one. The symbols in the right half of the table similarly refer to changes in the voting threshold t. Desirability criteria vary for different types of agents and will be outlined below. A transition is "neutral" if the two societies being compared are equivalent according to the chosen criterion. The '-' symbol (not to be confused with double neutrality '—') means "not applicable" and is used when there are no agents of the corresponding category.

For clarity, the diagrams are colored. The colors correspond to symbols or their combinations. They play a secondary role and are chosen informally. The combination in which ⊤ stands above ⊥ is called the *equilibrium 2-macrostate* and is marked in green like the equilibrium E.

3.2 Cooperation vs Atomization

What changes in society are beneficial to 1-agents, group members, and agents changing their strategy? The latter ones will be called *defectors*.

The appearance of a defector changes the group size g by one. This thus changes the arithmetic of the group members' strategy: they either begin to take into account the defector who joins the group, or stop taking into account the defector who leaves it.

A change in strategy is justified if it increases the defector's ECG. Testing this with respect to society S and society S_{+1} (whose group has one member more than in S) amounts to checking this condition for both possible transitions between S and S_{+1}. Namely, if the 1-agent's ECG in S is smaller than the group member's ECG in S_{+1}, then this 1-agent benefits from joining the group (which transforms S into S_{+1}) and an arrow pointing up to the S_{+1} cell is placed into the S cell. If, on the contrary, an 1-agent in S has a greater ECG than the group member in S_{+1}, then the latter agent benefits from leaving the group, and an arrow pointing to S is placed into the S_{+1} cell. These comparisons of ECG values are based on Corollary 1, for which the diagrams in Fig. 1 may be useful.

The S-cell contains one "vertical" symbol from the table in Fig. 2 that aggregates the results of evaluating transitions from S to both S_{+1} and S_{-1} (whose group has one less member than the group in S).

As already mentioned, the group size measures the level of cooperation in the society: $g = 0$ corresponds to complete atomization; a society with $g = n$ is fully cooperative.

The desirability diagram for defectors is shown in Fig. 2. Its arrows describe the processes that can occur on the initiative of defectors, provided that both entry into the group and exit from it are permitted. The main features of these processes are as follows.

If $t \in [5, 19]$, then 1-agents in societies with group size of 2 to 24 benefit from joining the group and group members are not interested in leaving. This opens the way for a "snowball" cooperation scenario [4,6]. When $g = 1$, an 1-agent benefits from joining a group, and nothing changes if the only group member becomes an 1-agent, since their group strategy is identical to the individualistic one. Every clique-like society ($g = 25$) is an equilibrium: leaving the group is unprofitable.

A general interpretation of this diagram is as follow: In the most liberal ($t < 5$) or conservative ($t > 19$) societies, the members of a moderate-sized group may have an incentive to atomize; in societies closer to majoritarianism (which means that decisions are made with a simple majority), 1-agents always have an incentive to choose cooperation. This is due to the fact that in liberal and conservative societies, the role of an individualist is highest: in the former, it is quite easy to implement her initiative, in the latter, a small faction can block changes that have high support. In some other cases we will also see that "extremes meet": in neutral environments, conservative and liberal societies have much in common that distinguishes them from majoritarian societies.

Conservatism in decision-making is "liberalism in reverse," that is, the freedom to block. As a result, the dominant unvarnished news agenda of some real-world conservative regimes is more about protests and reactions to them than about new ideas.

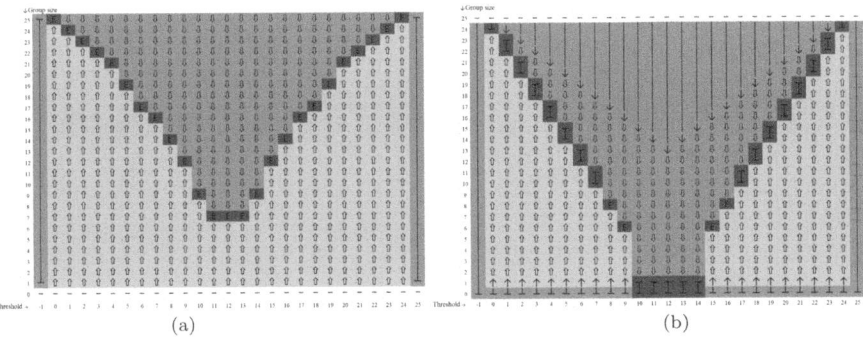

Fig. 3. Changes in the group size that benefit group members (a) or 1-agents (b). (Color figure online)

Let us now find out which changes in the structure of society are beneficial to agents who *do not change* their strategies: group members (Fig. 3a) or 1-agents (Fig. 3b).

In the case of 1-agents, the diagram has the following main regions:

- triangular area of neutrality (red), where all decisions are made by the group maximizing its capital gain, and the ECG of the 1-agents equals $\mu = 0$;
- Y-shaped area (yellow), where 1-agents benefit from increasing their number;
- two blue areas where the absolute difference between the voting threshold and the simple majority threshold (we will call this difference the *specificity* of a voting rule or a society) is 3 to 12, and the size of the group does not exceed twice the specificity. Here, 1-agents benefit from reducing their share in society.

Equilibrium states and 2-macrostates are marked in green. They maximize the ECG of 1-agents with respect to the group size for a fixed t.

The desirability diagram for group members (Fig. 3a) has a simple description: the closer the voting rule is to the majority rule, the smaller the optimal (for group members) group size; for $11 \leq t \leq 13$, this group size is 7. At each t, the group member's ECG is single-peaked.

Group members and 1-agents quite often benefit from the same changes in the degree of cooperation. In the diagram of Fig. 4d, a decrease (resp., an increase) in group size supported by these categories is marked in yellow (resp., in light blue); beige and dark blue colors respectively indicate a decrease and an increase with neutrality of one category and support of the other.

The areas of agreement of one of these categories and the defectors are shown in Figs. 4b and 4c; Fig. 4a presents the transitions approved by all agents.

All transitions of common agreement expand the group and thus fit into the "snowball of cooperation" scenario. This scenario has practical applications, among which we note [11], where it is described as follows: "The initial union [of actors] should be ready to accept new members, and those should wish to join

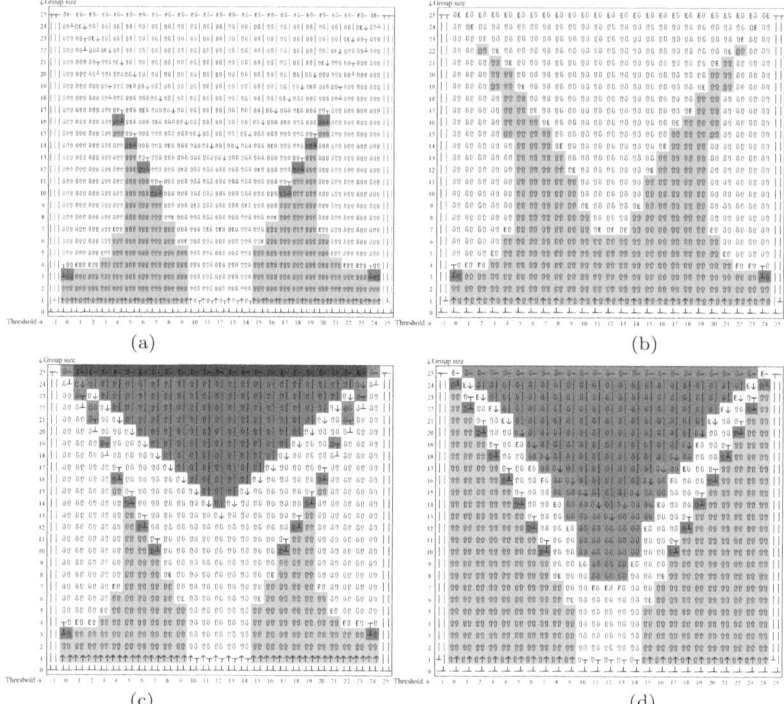

Fig. 4. Approval of transitions by: (a) a defector (the left symbol in each cell), group (the middle symbol), and 1-agents; (b) a defector (left symbol) and group; (c) a defector (left symbol) and 1-agents; (d) the group (left symbol) and 1-agents. (Color figure online)

this association. The reason for this desire can only be the obvious and significant advantages of the new status, and practical ones at that." In Fig. 4a, transitions approved by everyone are implemented at voting thresholds that differ from the majority threshold by at least 3, while the group size belongs to the interval [3, 16]. Transitions for which one of the categories is neutral are marked in dark blue. Together with the consensus transitions, they constitute 22% of non-trivial societies.

3.3 Democratic and Mafia Groups

So far we have assumed that each 1-agent can enter the group, and each group member can leave it by becoming an 1-agent. Such a group can be called *open*. However, many real-world groups, like political parties, behave differently. They do not accept everyone, but only those whose entrance seems beneficial for the group. Any member has the right to leave the group. In [11], a group of this type is described as follows: "Joining the association... should be conditioned by a meticulous assessment of the candidates' readiness, and exit should be unhindered."

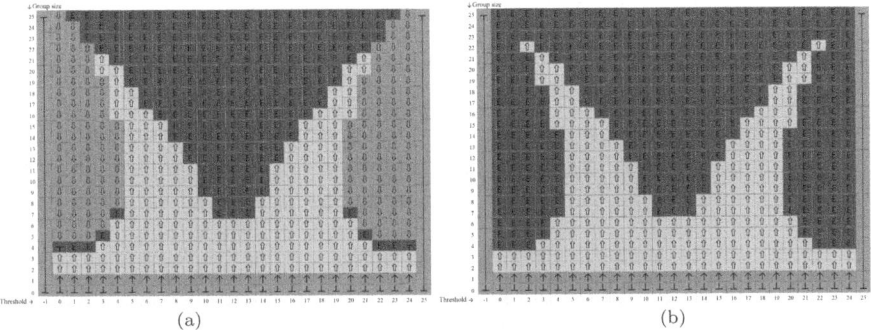

Fig. 5. Approved changes in the degree of cooperation in the presence of a democratic group (a) or a mafia group (b).

Figure 5a shows the transition diagram in the case where a necessary condition for joining the group, in addition to the benefit of its new member, is the consent of the group, which is given if the group expansion increases the ECG of the current group members. The justification for this conditioning in the ViSE model may be that when accepting a new member, the group must expand the concept of "group interest" by taking into account the interests of the defector, which reduces the degree of consideration of the interests of the current members.

Authorized entry and voluntary exit are consistent with the practice of democratic organizations, as opposed to mafia-type associations. Therefore, we will call this type of group *democratic*. If the ECG of the entering 1-agent remains the same, then her entry to the group will be considered approved if it increases the group members' ECG, prohibited if decreases, and neutral if preserves. This means the goodwill of the 1-agent: in the case of her own neutrality, she favors an action that benefits her possible partners. However, their benefit is lexicographically subordinate to the defector's own benefit.

With a democratic group, the more "specific" (conservative or liberal) the society, the higher the minimum level of cooperation that ensures its structural stability (equilibrium).

A diagram for a group that blocks *all* transitions that are unfavorable for its members (we call such a group a *mafia*) is presented in Fig. 5b.

The conclusion is that although a mafia group may be interested in reducing its size (Fig. 3a), it never approves of a member's initiative to leave it. Like a democratic group, it has the greatest prospects for expansion in moderately liberal or moderately conservative societies.

3.4 Liberalism vs Conservatism

We assume that the decision to change the voting threshold by one is made by voting with the current threshold. The change is supported by agents to which it brings an increase in the ECG. Diagrams of transitions consisting in changing the voting threshold are presented in Fig. 6.

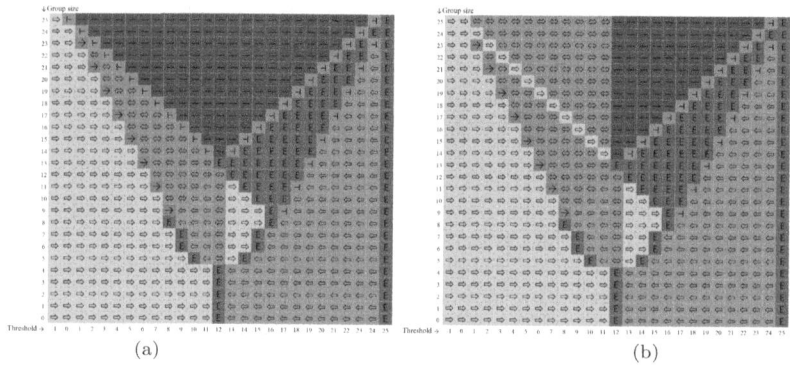

Fig. 6. Approved changes in the voting threshold made by voting with the current threshold. Abstention votes: (a) do not change support; (b) are counted in half. (Color figure online)

If a proposal preserves the agent's ECG, then this agent abstains; the votes of such agents do not contribute to the number of votes "for" (Fig. 6a) or are taken into account with a coefficient of 0.5 (Fig. 6b). In the latter case, an abstention provides 0.5 votes "for" and 0.5 votes "against". The corresponding diagrams have a minor difference: it concerns only the approval of certain transitions between the states equivalent in Fig. 6a. Let us consider the former rule.

The diagram in Fig. 6a shows that radically liberal or conservative societies tend to become less "specific". However, they become majoritarian only if the group size is less than 5. If the degree of cooperation is high, this movement soon ceases.

Interestingly, majoritarian and near-majoritarian societies with a moderate degree of cooperation ($g \in [5, 11]$) are characterized by a centrifugal tendency: they tend to increase their specificity. This is because of the individualists: a small proportion of them, together with the group, can accept a proposal that is advantageous to them in a relatively liberal society or reject an unfavorable proposal in a relatively conservative one; in a majoritarian society these possibilities are absent.

With greater cooperation, liberal societies fall into the red area of bidirectional transitions ("dynamic equilibria"), which then gives way to the green region of equilibrium macrostates, while conservative societies stop on the border of the light green area of equilibria.

3.5 Combinations of Structural and Procedural Transitions

The most interesting dynamics are realized when both structural and procedural transitions are allowed. Then, to analyze the possible evolution, we need to combine the diagrams shown in Figs. 2, 5a,b, and 6a. The results are shown in Fig. 7a (the case of an open group) and Fig. 7b (for a democratic group). The diagram for a mafia group is presented below in Fig. 9b. Since many of the 81 combinations of the transition symbols (Fig. 2) appear on the combined diagrams, the colors (which play a secondary role) only partially correspond to those attached to the symbols. The black stepped line shows the boundary of the Pareto set.

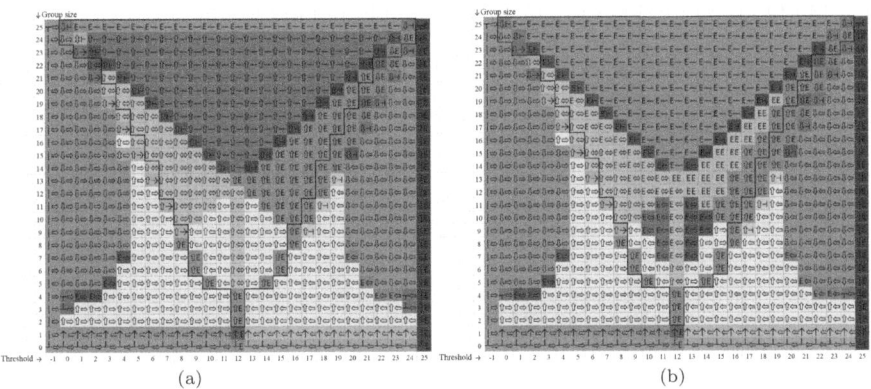

Fig. 7. Approved structural and procedural transitions in the case of: (a) an open group; (b) a democratic group. (Color figure online)

In the first case (Fig. 7a), there are two equilibrium macrostates: $t = 25$ (rejection of all proposals for any group size) and $g = 25$ for $1 \leq t \leq 23$ (all agents are in the group).

The main difference between the democratic and open group cases is the blocking of group's expansion (upward movement) in the large green triangle in Fig. 5a when the group is democratic. This leads to the appearance of (a) equilibrium macrostates at $14 \leq g \leq 24$ and (b) a region of equilibrium states, the corner points of which are $(12, 13)$, $(14, 11)$, and $(19, 19)$. For society as a whole, the case of a democratic group is often less beneficial: by prohibiting the unprofitable entry of 1-agents, the group limits the growth of society's capital.

4 Combined Evolution Scenarios

By *combined scenarios* (routes) we mean the scenarios that include changes in both the structure of society and the voting threshold, where all changes are approved using the relevant rules discussed above. Can such scenarios be cyclic?

The answer is yes; a dark blue counterclockwise cycle is shown in Fig. 8a. All states in the region bounded by this cycle plus the $(2, 22)$ state are mutually accessible; the corresponding societies are majoritarian or liberal. This trapezoidal region is described as $t \in [2, 12]$, $g \in [4, 22 - t]$. It is accessible from different parts of the 'universe'. Exceptions are some conservative societies and the large green upper triangle.

Note, however, that conservative societies with low cooperation ($g \in [0, 4]$; recall that $g = 0$ and $g = 1$ are equivalent) or extreme conservatism ($t \in [22, 24]$) without extremely large ($g > 22$ for $t = 22$ or $g = 25$ for $t = 23$) groups have paths to that trapezoid. Typical evolution paths are shown by black arrows.

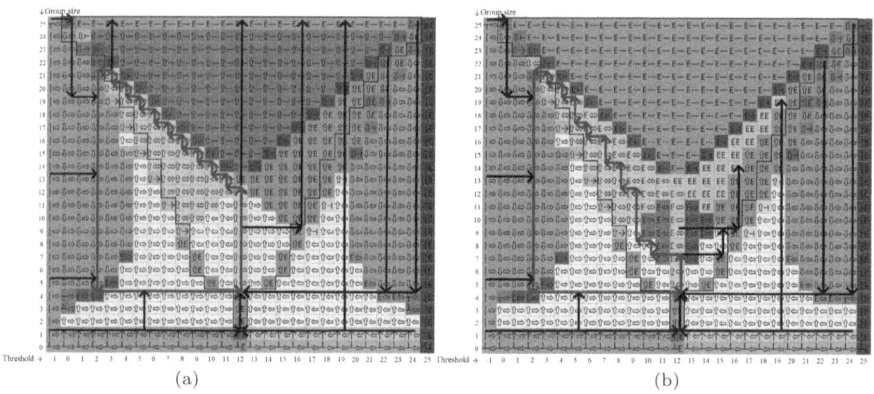

Fig. 8. Combined scenarios with: (a) an open group; (b) a democratic group. (Color figure online)

How does the ECG of society change along the considered cyclic route? Its dependence on the step number starting from $(t = 2, g = 21)$ is shown in Fig. 9a. Along this cycle, any increase in atomization (yellow) reduces the ECG, any move toward majoritarianism (red) increases it, while the impact of increased cooperation (blue) or increased liberalism (green) is ambiguous.

In the case of a democratic group (Fig. 8b), some 'upward' transitions are prohibited by the group. A counterclockwise cycle still exists, but the region of mutually accessible states is smaller.

Mafia groups do not allow their reduction at the initiative of the participants, so no cyclic scenarios are possible. However, there are scenarios with a non-monotonic change in the threshold, one of which is shown in Fig. 9b.

5 Conclusion

Within the framework of the ViSE model, we studied the evolution of a two-component society, which consists of a change in the level of cooperation (group

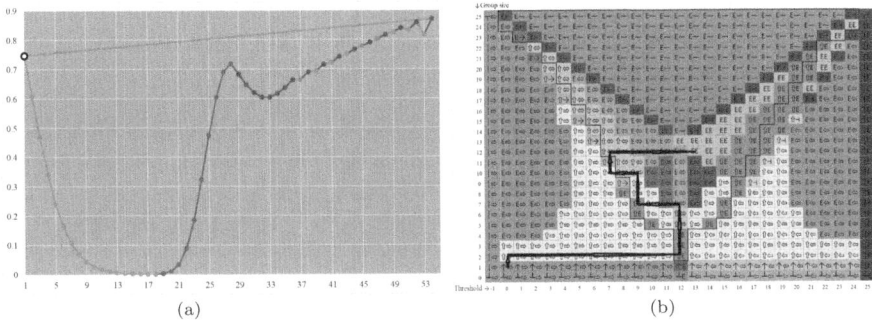

Fig. 9. (a) ECG of the societies on the cyclic route shown in Fig. 8a depending on the step number starting from $(t = 2, g = 21)$ at $\sigma = 12$; growth of: atomization (yellow); cooperation (blue); liberalism (green); approaching majoritarianism (red). (b) A combined scenario for societies with a mafia group. (Color figure online)

size g) and the degree of conservatism (voting threshold t). We considered societies with $n = 25$ members; the main qualitative findings hold for increasing or moderately decreasing n.

Non-trivial evolution patterns were identified. In the case of an open group, all societies belonging to the trapezoid $t \in [2, 12]$, $g \in [4, 22 - t]$ participate in cyclical patterns with all other societies of this region. They are accessible from conservative societies with low cooperation or very high conservatism without extremely large groups and from other liberal or majoritarian societies not belonging to the upper green triangle in Fig. 8a. For societies with democratic groups, cyclical patterns exist, but some transitions that expand the group are prohibited. A mafia group never lets its members go when they want to, but there are patterns with a non-monotonic change in the threshold t.

The insights derived from parsimonious models, such as the ViSE framework, facilitate the elucidation of conditions that potentially govern the occurrence of similar phenomena in empirical contexts.

Acknowledgment. This work was supported by the Israel Science Foundation (grant No. 1225/20) and by the European Union (ERC, GENERALIZATION, 101039692).

References

1. Afonkin, V.A.: Tax incentives for prosocial voting in a stochastic environment. Control Sci. **1**(1), 53–59 (2021). https://doi.org/10.25728/cs.2021.1.6
2. Aizerman, M.A.: Dynamic aspects of voting theory (survey). Autom. Remote. Control. **42**(12), 1664–1675 (1981)
3. Chebotarev, P., Afonkin, V.: Majority voting is not good for heaven or hell, with mirrored performance. Ann. Oper. Res. (2025)

4. Chebotarev, P., Maksimov, V.: Two-component societies in the ViSE model: how the level of cooperation and voting threshold affect the capital dynamics. Large-Scale Syst. Control **93**, 51–88 (2021). https://doi.org/10.25728/ubs.2021.93.2. (in Russian)
5. Chebotarev, P.Y.: Analytical expression of the expected values of capital at voting in the stochastic environment. Autom. Remote Control. **67**(3), 480–492 (2006). https://doi.org/10.1134/S000511790603012X
6. Chebotarev, P.Y., Loginov, A.K., Tsodikova, Y.Y., Lezina, Z.M., Borzenko, V.I.: Analysis of collectivism and egoism phenomena within the context of social welfare. Autom. Remote Control. **71**(6), 1196–1207 (2010). https://doi.org/10.1134/S0005117910060202
7. Chebotarev, P.Y., Malyshev, V.A., Tsodikova, Y.Y., Loginov, A.K., Lezina, Z.M., Afonkin, V.A.: The optimal majority threshold as a function of the variation coefficient of the environment. Autom. Remote Control. **79**(4), 725–736 (2018). https://doi.org/10.1134/S0005117918040136
8. Chebotarev, P.Y., Tsodikova, Y.Y., Loginov, A.K., Lezina, Z.M., Afonkin, V.A., Malyshev, V.A.: Comparative efficiency of altruism and egoism as voting strategies in stochastic environment. Autom. Remote Control. **79**(11), 2052–2072 (2018). https://doi.org/10.1134/S0005117918110097
9. Enelow, J.M., Hinich, M.J.: The Spatial Theory of Voting: An Introduction. Cambridge University Press, Cambridge (1984)
10. Galasso, V., Profeta, P.: The political economy of social security: a survey. Eur. J. Polit. Econ. **18**(1), 1–29 (2002). https://doi.org/10.1016/S0176-2680(01)00066-0
11. Kovalev, S.A.: Civic responsibility of intellectuals (Political idealism and real politics: The challenge of the 21st century). Novaya Gazeta (2010). https://novayagazeta.ru/articles/2010/03/12/4341-grazhdanskaya-otvetstvennost-intellektualov
12. Kranich, L.: Altruism and the political economy of income taxation. J. Public Econ. Theory **3**(4), 455–469 (2001). https://doi.org/10.1111/1097-3923.00078
13. Levine, D.K.: Modeling altruism and spitefulness in experiments. Rev. Econ. Dyn. **1**(3), 593–622 (1998). https://doi.org/10.1006/redy.1998.0023
14. Lindenberg, S.: Social rationality versus rational egoism. In: Turner, J.H. (ed.) Handbook of Sociological Theory, vol. 29, pp. 635–668. Kluwer Academic/Plenum Publisher, New York (2001)
15. Maksimov, V.M., Chebotarev, P.Y.: Voting originated social dynamics: quartile analysis of stochastic environment peculiarities. Autom. Remote Control. **81**(10), 1865–1883 (2020). https://doi.org/10.1134/S0005117920100069
16. Malyshev, V.: Optimal majority threshold in a stochastic environment. Group Decis. Negot. **30**(2), 427–446 (2021). https://doi.org/10.1007/s10726-020-09717-8
17. Margolis, H.: Selfishness, Altruism, and Rationality. University of Chicago Press, Chicago (1984)
18. McKelvey, R.D.: Intransitivities in multidimensional voting models and some implications for agenda control. J. Econ. Theory **12**(3), 472–482 (1976). https://doi.org/10.1016/0022-0531(76)90040-5
19. Mirkin, B.G.: Group Choice. V.H. Winston & Sons, Washington D.C. (distributed by Halsted Press Division of John Wiley & Sons, N.Y.) (1979)
20. Pavlova, A., Rigobon, R.: An asset-pricing view of external adjustment. J. Int. Econ. **80**(1), 144–156 (2010). https://doi.org/10.1016/j.jinteco.2009.09.003
21. Roberts, K.: Voting over income tax schedules. J. Public Econ. **8**(3), 329–340 (1977). https://doi.org/10.1016/0047-2727(77)90005-6

22. Romer, T.: Individual welfare, majority voting, and the properties of a linear income tax. J. Public Econ. **4**(2), 163–185 (1975). https://doi.org/10.1016/0047-2727(75)90016-X
23. Tsodikova, Y., Chebotarev, P., Loginov, A.: Modeling responsible elite. In: Aleskerov, F., Vasin, A. (eds.) Recent Advances of the Russian Operations Research Society, vol. 6, pp. 89–110. Cambridge Scholars Publishing, Newcastle upon Tyne (2020)
24. Tsodikova, Y.Y., Chebotarev, P.Y.: Modeling society with a responsible elite. J. New Econ. Assoc. (1(66)), 12–35 (2025). https://doi.org/10.31737/22212264_2025_1_12-35. (in Russian)
25. Weron, A., Weron, R.: Inżynieria Finansowa. Wydawnictwo Naukowo-Techniczne, Warszawa (1998)

Preference Modeling, Evaluation, and Decision Support in Group Contexts

Selection of Digital Technologies for Energy Management: A Group Decision Approach Based on PROMETHEE-ROC

Pedro Henrique Gouveia de Souza[1], Nathália Jucá Monteiro[2], Sergio Eduardo Gouvea da Costa[3], and Eduarda Asfora Frej[1](✉)

[1] Universidade Federal de Pernambuco – UFPE, CDSID - Center for Decision Systems and Information Development, Av. Acadêmico Hélio Ramos, s/n – Cidade Universitária, Recife, PE 50.740-530, Brazil
eafrej@cdsid.org.br

[2] Universidade do Estado do Pará, Marabá, Pará, Brazil

[3] Universidade Tecnológica Federal do Paraná, Av. Sete de Setembro 3165, Curitiba, Paraná, Brazil

Abstract. Energy consumption is a critical issue that affects countries and communities worldwide. Its efficient use is essential for economic development, environmental preservation, and the well-being of societies. In this scenario, the manufacturing sector stands out as a major consumer of energy. In Brazil, manufacturing industry is the largest consumer of energy. Hence, energy management (EM) should be applied to minimize energy costs and/or reduce energy consumption. Monitoring the system's consumption of power is the first step towards reducing the consumption of energy. Digital technologies (DTs) can help achieve this monitoring. DTs are triggering a revolution in energy areas and are being used in energy system planning, design, control, and optimization. Energy related problems inherently involve multiple and conflicting criteria, as well as the perceptions of multiple stakeholders. In this context, this paper aims to address the issue of energy management policies in Brazilian companies, proposing a multicriteria group decision making (MCGDM) model for prioritizing DTs for EM. The PROMETHEE-ROC methodological structure was adapted for a group decision context and applied for ranking 14 DTs selected considering the Brazilian context. The model developed seeks to prioritize/select DTs that help achieve greater energy efficiency, thereby improving energy management in Brazilian industries.

Keywords: Energy management · Digital Technologies · Multicriteria Group Decision Making (MCGDM) · PROMETHEE-ROC

1 Introduction

Energy is a worldwide concern. In 2022, the world invested about 453 billion dollars in energy efficiency, compared to the 343 billion dollars that it invested in 2015 (International Energy Agency 2023). Also in 2022, the United Nations set the 7th of the

Sustainable Development Goals (SDG) for 2030, which is directly related to energy efficiency and the use of renewable energies (United Nations 2023). For industry, this concern is even greater due to rising power costs (Adenuga et al. 2019) and because industrial parks are major consumers of energy (Anastasovski 2023).

Brazil is one of the world's largest energy consumers, and the manufacturing sector is the largest consumer of electricity among all the sectors of production (Energy Research Company 2023). Hence, energy management (EM) is a fundamental task for Brazilian industries, in order to seek how best to minimize energy costs and/or to reduce energy consumption. Recent research has shown that efficient EM brings a positive commercial impact for industries, and leads to reducing costs and generating other benefits (Roth et al. 2020; Sivill et al. 2013). In this context, Digital Technologies (DTs) can help monitor and analyze energy data, thus helping to make the process more efficient and to avoid losses, and hence contributing to EM.

Since energy problems involve multiple objectives, multicriteria decision-making (MCDM) methods are useful tools for dealing with these problems (Mardani et al. 2017). MCDM methods have already been used to select energy projects (Kshanh and Tanaka 2024) or energy equipment (Hosouli 2024), and also to prioritize activities for energy efficiency (Richter et al. 2023). DTs are already associated with MCDM methods, such as selecting the best technology for industrial plants (Maretto et al. 2022) or in circular supply chains (Tanveer et al. 2023). Energy related problems also involve multiple stakeholders with distinct preferences and/or backgrounds and they have an impact on setting and meeting the multiple objectives and alternatives of the problem.

In this context, this paper proposes a Multicriteria Group Decision Making (MCGDM) model to rank DTs applied to energy management problems in Brazilian industries. 14 DTs that are often associated with energy management systems in industrial environments have been pre-selected to be evaluated. 17 criteria were drawn up based on a literature review, and the opinions of three decision-makers who specialize in digital technologies were considered in the evaluation. The PROMETHEE-ROC (Morais et al. 2015) is applied for ranking alternatives; this method was chosen due to its simplicity for the DMs considering the low amount of information to be provided, as well as its easiness of use and straightforward adaptation for dealing with group decision problems. The application of PROMETHEE-ROC was conducted in two stages: i) first, an individual evaluation of the alternatives is conducted based on the methodological structure of PROMETHEE-ROC; ii) then, individual results are aggregated considering the methodological foundations of the PROMETHEE method for group decision situations (Mareschal et al. 1998). The application is followed by a discussion on further implications with regard to issues that arise when implementing DTs.

This paper is structured as follows. Section 2 presents a brief background on the PROMETHEE-ROC method and its methodological adaptation for dealing with group decision problems. Section 3 describes the problem of prioritizing DTs. Section 4 presents the results obtained by applying the PROMETHEE multicriteria method, and finally Sect. 5 discusses the results obtained and draws final conclusions.

2 Group Decision Modeling with PROMETHEE-ROC

2.1 PROMETHEE-ROC Method

The PROMETHEE-ROC method (Morais et al. 2015) is a variation of the traditional PROMETHEE (Preference Ranking Organization Method for Preference Ranking Organization Method for Enrichment Evaluations), a family of methods of multiple criteria decision-making (MCDM) for decision-makers (DMs) with non-compensatory rationality (Vansnick 1986). Its procedure is based on pairwise comparisons between alternatives and has two phases (Brans and Vincke 1985): constructing outranking relationships (I) and exploiting these relationships for decision recommendations (II).

In the first phase, outranking relationships are defined from a pairwise comparison between alternatives. In PROMETHEE, these outranking relations are valued by an outranking degree ($\pi(a, b)$), which measures the strength with which each alternative a outperforms alternative b. In order to compute the outranking degree, a preference function P_j should be defined for each criterion $j (j = 1, \ldots, n)$; this function simply indicates the intensity of preference of alternative a over alternative b in criterion, $P_j(a, b)$, taking into account the difference in performance of the two alternatives in this criterion.

The preference function can be defined as the way in which the strength of the DM's preference increases with the increasing difference between the performance of the alternatives on a given criterion (de Almeida et al. 2015). Preference functions can assume six basic forms (see Brans and Vincke (1985) for details), which may require indifference and/or preference thresholds to be defined, depending on the function chosen by the DM.

The outranking degree $\pi(a, b)$ is calculated based on pairwise comparisons between alternatives made by the DM, by using the preference function $P_j(a, b)$ weighted by the importance level of each criterion, w_j. The outranking degree of alternative a over alternative b is calculated using Eq. (1), and weights are normalized and sum up to 1, as shown in Eq. (2).

$$\pi(a, b) = \sum_{j=1}^{n} P_j(a, b) \times w_j \quad (1)$$

$$\sum_{j=1}^{n} w_j = 1 \quad (2)$$

So, once the outranking degrees have been established, for PROMETHEE II, the positive outranking flow (φ^+) and the negative outranking flow (φ^-) of each alternative must be calculated in order to obtain the net flow (φ) and thus to rank the alternatives. The positive outranking flow can be understood as the intensity of preference of alternative a over the other alternatives, while the negative outranking flow is the counterpart of the other alternatives in relation to alternative a. The outranking flows can be calculated using the following equations:

$$\varphi^+(a) = \frac{\sum_{b \in A}^{n} \pi(a, b)}{n - 1} \quad (3)$$

$$\varphi^-(a) = \frac{\sum_{b \in A}^{n} \pi(b, a)}{n - 1} \quad (4)$$

$$\varphi(a) = \varphi^+(a) - \varphi^-(a) \tag{5}$$

Hence, the net flow can be understood as the balance relative to the intensity of preferences of alternative a. Based on the net flow, the alternatives can be ranked in descending order of φ, thereby obtaining a complete order of the alternatives.

An issue related to outranking methods concerns setting the parameters of the model, especially the criteria weights. It may be difficult for DMs to define exactly the degree of importance of each criterion. The use of surrogate weighting methods can be very useful in these situations (Da Silva et al. 2023).

Morais et al. (2015) developed a variant of the PROMETHEE II method in which weights are calculated according to the ROC (Rank Order Centroid) surrogate weighting method (Barron and Barrett 1996). In the so-called PROMETHEE-ROC, the only information required from the DM is the order of the criteria ($w_1 > w_2 > \ldots > w_j > \ldots > w_n$), and then the weight of criterion j is calculated according to the position it occupies in the ranking, based on Eq. (6). In numerical terms, the ROC procedure assigns the highest weight to the criterion selected as being the most important and the process continues for all the positions in the ranking defined by the DM (Barron 1992).

$$w_j = \frac{1}{n} \sum_{k=j}^{n} \frac{1}{k}, j = 1, 2, \ldots, n. \tag{6}$$

Based on PROMETHEE-ROC, a decision support system (DSS) called PROMETHEE *Roc n Ratio* has been launched in recent years, offering decision support based on PROMETHEE II, with the weights being elicited from the ROC, using the Ratio procedure or by defining the weights manually. This DSS is available free of charge at: http://cdsid.org.br/prometheeroc/.

Regarding the ROC application for the PROMETHEE method, one could wonder why not using other surrogate weighting procedures such as Rank-Sum (RS), Reciprocal of Ranks (RR), or Rank Exponential weights (Stillwell et al. 1981). In fact, the literature shows that ROC outperforms other surrogate methods when applied in the structure of the PROMETHEE method. In the work of de Almeida Filho et al. (2018), experiments were conducted applying surrogate weighting procedures to analyze their performance in PROMETHEE method. The findings show that, for the choice problem, the ROC procedure showed better correspondence with the weights used in PROMETHEE based on metrics such as the highest Hit Rate and lowest Average Loss Value (de Almeida Filho et al. 2018). As for the ranking problems, the ranking obtained with each substitute procedure was compared with the "true ranking" using the Kendall test, so that the ROC was the surrogate weighting procedure with the highest correlation with the "true" ranking. Thus, all the results indicate that ROC is the most suitable surrogate weighting procedure to be used with the PROMETHEE II method for choice and ranking problematics (de Almeida Filho et al., 2018).

2.2 Group Decision with PROMETHEE

In group decision problems, individual evaluations in PROMETHEE can be aggregated to find the group ranking of the alternatives (Mareschal et al. 1998). The aggregation is conducted based on the individual net flows obtained by each DM, taking into account the degree of importance of each DM to the decision problem (Mareschal et al. 1998). This ensures that each DM has actively participated in the decision problem, so that the overall recommendation reflects their preference structure to some extent (see Fig. 1).

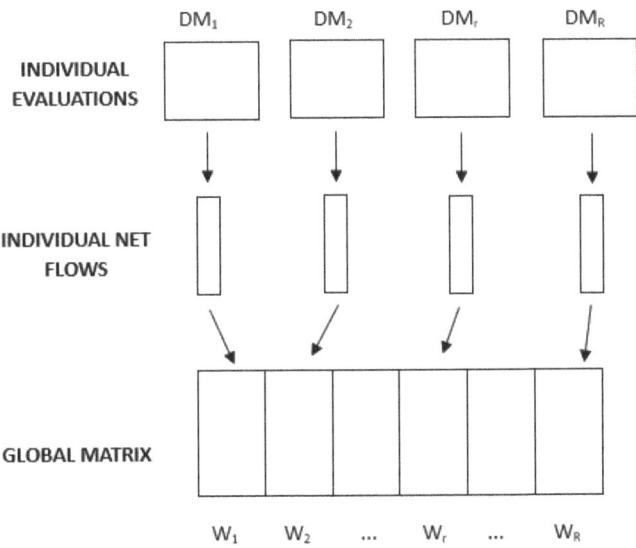

Fig. 1. Incorporation of individual evaluations into a global evaluation matrix (adapted from Mareschal et al. 1998).

Once the individual net flows have been found, the overall net flow of each alternative can be defined ($\varnothing^G$) as the weighted sum of the individual net flows of each alternative by the weight W_r of each decision-maker $r (r = 1, \ldots, R)$, as shown in Eq. (7). DMs' weights are normalized and sum up to 1, as seen in Eq. (8).

$$\varnothing^G(a_j) = \sum_{r=1}^{R} \varphi^r(a_j) \times W_r \quad (7)$$

$$\sum_{r=1}^{R} W_r = 1 \quad (8)$$

Hence, a group order of the alternatives can be obtained by ranking alternatives in descending order of their overall net flows $\varnothing^G$. Note that the use of conflict resolution tools and the application of sensitivity analysis is important to provide greater certainty about the overall recommendation (Mareschal et al. 1998).

3 Digital Technologies for Energy Management: A Prioritization Problem

3.1 Description of the Problem

The decision problem aims to generate a prioritization order for different DTs, with a view to optimizing energy management for Brazilian companies. The problem is situated in the group decision area because prioritizing these digital technologies involves multidisciplinary areas, so the participation of different actors with diverse backgrounds and/or preferences can contribute significantly to the decision-making process. Thus, the problem can be evaluated more robustly, as the analysis of different DMs brings different perspectives from which the decision problem can be approached.

Several applications have been made in recent years on the subject of energy efficiency, in order to associate the application of different DTs with benefits in the energy sector. In order to define the alternatives and criteria for the decision problem, a literature review was carried out on the subject, in addition to consulting and involving different experts during the decision-making process (Frank et al. 2019; Pessl et al. 2020; Schumacher et al. 2016; Lyu and Liu 2021; Maroufkhani et al. 2022).

To select the 17 criteria, a systematic literature review was conducted across three databases (Science Direct, Web of Science, and Scopus), focusing on publications from the last five years (2018–2023). A total of 40 papers were selected and analyzed to identify the criteria and the DT. All selected criteria are considered relevant for the

Table 1. Definition of the Criteria.

Criterion	Definition
C1- FINAIC	Financial incentives received for adopting digital transformation technologies
C2- COPRO	Cost reduction in existing processes
C3- COINS	Installation costs of digital transformation technologies
C4- COMAIN	Maintenance costs for digital transformation technologies
C5- DURAB	Expected lifespan of the technology
C6- SCALAB	Ability to expand technology in industrial operations as they grow or increase in complexity
C7- RELIAB	Ability of the technology to be robust and able to perform as expected in different conditions and scenarios
C8- EASUS	Simplicity and accessibility to use the technologies implemented
C9- EASIMP	Simplicity and accessibility to implement the technologies
C10- FLEXB	Ability of the technology to be used in flexible or autonomous production lines
C11- EFFITI	Ability to analyze data in real time and make decisions more quickly
C12- EFFIRES	Ability to optimize industrial processes through intelligent automation
C13- CENTRAL	Capacity to centralize and coordinate the various processes and systems in an industrial environment
C14- NOISE	Unwanted variations and inaccuracies in the signal collected by the sensors used to monitor and collect process data
C15- SUSSOC	Capacity to create specialized jobs
C16- SUSECO	Capacity to reduce energy consumption
C17- SUSENV	Capacity to reduce carbon emissions and use energy resources efficiently

context of Brazilian energy management, because they address economic, technical, and sustainability aspects that can impact industries, especially Brazilian ones dealing with high energy prices and significant pollutant energy sources.

A total of 17 criteria were defined based on economic, environmental, technical and strategic aspects. Table 1 details the criteria used.

The 14 technologies encompass software and hardware used in energy management problems. The DTs are described in Table 2.

Table 2. List of digital technologies (DTs) to be ranked

DT	Description
Internet of Things (IoT)	Interconnection of physical devices, machines, and objects through the internet
Cyber-Physical Systems (CPS)	Cybernetic and physical subsystems that assist in the collection, transmission, and analysis of data
Industrial Wireless Networks (IWN)	Communication systems that enable connectivity among devices, machines, and systems in industrial environments through wireless technologies
Edge Computing (EC)	An approach that involves data processing and analysis happening closer to where the data is generated, instead of sending it to remote processing centers
Cloud Computing (CC)	The hardware, software, and data resources are stored and accessed remotely through the internet
Cybernetic Systems (CyberS)	They are interconnected systems that involve the interaction among mechanical, electronic, and software components, along with constant feedback to control and regulate performance
Communication Systems (CommunS)	Machine-to-machine communication
Sensors	Devices for monitoring systems (some in real-time)
Digital Twin (DT)	Virtual replicas of physical assets, processes, or systems
Autonomous robots/systems (ARS)	Sets of technologies that operate independently
Artificial Intelligence (AI)	Creation of systems that can simulate human thought processes, with or without explicit programming
Big Data	Sets involving the collection, storage, and analysis of complex and varied information in large volumes
Blockchain	Distributed ledger technology that creates an interlinked chain of blocks to store information securely and transparently
3D Printing	Technology that enables the creation of three-dimensional objects from digital models using successive layers of materials such as plastic, metal, or ceramic

Constructed scales were used to evaluate each alternative in each criterion. Most of them consisted of a 5-point Likert scale from 1 to 5, but some were built from 1 to 3 (3-point Likert scale) since a significant level of specification was optional.

Three decision makers were interviewed. All the participants have experience working and/or researching in Brazilian industries. They used to act as external consultants in manufacturing projects. The first decision-maker (DM1) has experience in digital transformation and manufacturing systems and has also been involved in project coordination, software development, and research. The second DM (DM2) has been working with digital technologies in healthcare for six years. The third DM (D3) has five years of experience in Industry 4.0 and has used digital technologies such as artificial intelligence, big data, and business intelligence for manufacturing. The DMs were chosen based on their experience with digital technologies.

The process of collecting DMs' information started with a first contact by e-mail. The work and process were explained, and after their acceptance, an Excel sheet containing all the criteria, their explanations, and the DT with descriptions was sent to them. All instructions for completing the sheet were provided in the email. Essentially, the process was conducted accordingly online.

Since all criteria were measured in constructed scales mostly with qualitative meaning, an outranking method was considered as more appropriate for conducting the analysis, since it would be hard for DMs to perform tradeoffs between qualitative options in different criteria, if an additive aggregation compensatory model were to be applied. Moreover, a non-compensatory rationality is more suitable for this problem, since the selected technologies should be minimally acceptable for all criteria, without compensation possibilities. Hence, the PROMETHEE multicriteria method was chosen to be applied in this model, due to its non-compensatory characteristic and because it is easy to adapt it to deal with group decision problems (Mareschal et al. 1998).

4 Prioritizing Digital Technologies with PROMETHEE-ROC

A variant of the PROMETHEE method introduced in Sect. 2, the PROMETHEE-ROC, was applied to each DM individually; they evaluated the problem and ordered the criteria of the problem according to the degree of importance they considered, so that the value of the weights of each criterion was calculated via ROC (Rank Order Centroid).

Considering a non-compensatory approach, the PROMETHEE-ROC multicriteria method was chosen to be applied in this case due to its facility of use and simplicity. A minimum amount of information is required from decision makers (only the rank of criteria weights), and the availability of a free Decision Support System (DSS) also facilitates the application of the method.

For simplification purposes, the usual preference function was used for all the criteria for each DM. With the proposal to adapt the method to group decision-making, once the individual net flows had been found, the overall net flow was calculated, considering the weight of each decision-maker. For this particular problem, it was considered that the three DMs have the same degree of importance to the problem, since they are specialists in different areas, and each brings an important perspective to the decision.

Each DM individually ordered the criteria of the problem. Then, the weights of the criteria were calculated by the ROC. For DM1, the order of importance of the criteria was as follows: C4 > C5 > C12 > C15 > C3 > C11 > C2 > C1 > C6 > C13 > C17 > C7 > C9 > C14 > C8 > C10 > C16.

In terms of ranking, the result was that the highest priority technology is Big Data, followed by Artificial Intelligence, IoT and Blockchain. Table 3 details the ranking, as well as the respective flows.

Table 3. Ranking for DM1.

Alternatives	Positive Flow, $\varphi +$	Negative Flow, $\varphi -$	Net Flow, $\varphi °$	Ranking
Big Data	0.5014	0.0885	0.4129	1
Artificial Intelligence – AI	0.5118	0.1342	0.3776	2
IoT	0.4611	0.1175	0.3436	3
Blockchain	0,4517	0.1595	0.2921	4
Sensor	0.3888	0.2412	0.1476	5
Communication systems – CommunS	0.3739	0.2363	0.1375	6
Cloud computing – CC	0.3642	0.2484	0.1158	7
Edge Computing – EC	0.3545	0.3331	0.0214	8
3D Printing	0.2661	0.4367	−0.1706	9
Autonomous robots/systems – ARS	0,2204	0.4834	−0.263	10
Digital Twins – DT	0. 238	0.5175	−0.2795	11
Cyber-Physical Systems – CPS	0.1829	0.4971	−0.3143	12
Cybernetic systems – CyberS	0.1734	0.5022	−0.3287	13
Industrial wireless networks – IWN	0.1351	0.6277	−0.4926	14

For the second DM, the order of importance of the criteria was as follows: C6 > C7 > C8 > C15 > C4 > C2 > C5 > C9 > C16 > C17 > C10 > C3 > C13 > C1 > C11 > C12 > C14. As a result, the alternative Big Data also came top of the ranking, followed by Artificial Intelligence, Cloud Computing and IoT. Table 4 details the ranking for DM2.

Table 4. Ranking for DM2.

Alternatives	Positive Flow, $\varphi+$	Negative Flow, $\varphi-$	Net Flow, $\varphi°$	Ranking
Big Data	0.5560	0.0889	0.4671	1
Artificial Intelligence – AI	0.5573	0.0942	0.4632	2
Cloud computing – CC	0.5317	0.1189	0.4127	3
IoT	0.4628	0,1155	0.3473	4
Edge Computing – EC	0.3494	0.2437	0.1057	5
Communication systems – CommunS	0.3263	0.2808	0.0455	6
Blockchain	0.2802	0.3138	−0.0336	7
Sensor	0.2966	0.3585	−0.0619	8
Cyber-Physical Systems – CPS	0.2394	0.3620	−0.1226	9
Cybernetic systems – CyberS	0.1895	0.4079	−0.2184	10
Autonomous robots/systems – ARS	0.205	0.4934	−0.2884	11
3D Printing	0.1567	0.5041	−0.3474	12
Digital Twins – DT	0.2046	0.5692	−0.3646	13
Industrial wireless networks – IWN	0.1458	0.5504	−0.4046	14

For the third DM the order of importance of the criteria was as follows: C15 > C16 > C17 > C6 > C7 > C11 > C12 > C13 > C14 > C10 > C5 > C9 > C8 > C1 > C4 > C3 > C2. Thus, the alternative Artificial Intelligence was placed first in the ranking, and it was followed by Clouding Computing, Big Data and Edge Computing. Table 5 shows the details of the ranking for DM3.

Finally, from the individual net flow of each DM, the overall net flow of the problem could be calculated, taking into account that, for this problem, the three DMs were deemed to have equal weight, since each of them is a specialist in a different area. Thus, the overall ranking could be obtained, taking into account the context of the group decision. Table 7 shows the overall ranking obtained by using the global net flow.

It can be seen that the Artificial Intelligence alternative was defined as the problem's highest priority technology, although it was only DM3 who placed it first in the ranking. Moreover, the Big Data alternative came second in the global ranking, even though it was the highest priority for DM1 and DM2. This is due to the influence of net flow on the ranking. This example shows how important it is to make a decision based on solid numbers, rather than on the average position in the ranking, for example. The overall net

Table 5. Ranking for DM3.

Alternatives	Positive Flow, $\varphi +$	Negative Flow, $\varphi -$	Net Flow, $\varphi °$	Ranking
Artificial Intelligence – AI	0.5317	0.029	0.5026	1
Cloud computing – CC	0.4732	0.0888	0.3844	2
Big Data	0.3996	0,1789	0.2207	3
Edge Computing – EC	0.3936	0.1775	0.2161	4
IoT	0.3539	0.1452	0.2087	5
Communication systems – CommunS	0.3065	0.208	0.0985	6
Blockchain	0.3199	0.2605	0.0594	7
Cybernetic systems – CyberS	0.2743	0.3175	−0.0431	8
Sensor	0.264	0,3265	−0.0625	9
Digital Twins – DT	0.2387	0.4451	−0.2064	10
Cyber-Physical Systems – CPS	0.2092	0.4163	−0.2072	11
Autonomous robots/systems – ARS	0.2156	0.4254	−0.2099	12
Industrial wireless networks – IWN	0.1334	0.5506	−0.4173	13
3D Printing	0.1025	0.6466	−0.5441	14

Note that the ranking of the criteria was the only preferred information given by the three DMs. The weight of each criterion for each DM was calculated via ROC. This avoids the bias that could arise if a DM placed too high a weight on a specific criterion. Table 6 details the weight of each criterion for the three DMs

flow directly reflects the group's preference structure, taking into account the degree of importance that each DM has for the problem.

At the end of the process, a sensitivity analysis was carried out for the preference of each individual DM, based on Monte Carlo simulation. Thus, for each DM, the consequences and weights for each criterion were varied by 10%. For the three DMs, a robustness (percentage of cases in which an alternative remained in the same position) of approximately 60% was observed for the alternatives in the top four positions of each DM ranking. Thus, when considering the global ranking, although there may be a few changes between the first four positions, the global ranking can be considered robust.

Table 6. Weights of criteria for DMs

Criterion	DM1 -w_j	DM2 -w_j	DM3 -w_j
C1- FINAIC	0.0498	0.0153	0.0153
C2- COPRO	0.0582	0.0680	0.0035
C3- COINS	0.0798	0.0247	0.0071
C4- COMAIN	0.2023	0.0798	0.0111
C5- DURAB	0.1435	0.0582	0.0300
C6- SCALAB	0.0425	0.2023	0.0945
C7- RELIAB	0.0247	0.1435	0.0798
C8- EASUS	0.0111	0.1141	0.0198
C9- EASIMP	0.0198	0.0498	0.0247
C10- FLEXB	0.0071	0.0300	0.0359
C11- EFFITI	0.0680	0.0111	0.0680
C12- EFFIRES	0.1141	0.0071	0.0582
C13- CENTRAL	0.0359	0.0198	0.0498
C14- NOISE	0.0153	0.0035	0.0425
C15- SUSSOC	0.0945	0.0945	0.2023
C16- SUSECO	0.0035	0.0425	0.1435
C17- SUSENV	0.0300	0.0359	0.1141

Table 7. Global Ranking for the decision problem.

Alternatives	DM1 φ	DM2 φ	DM3 φ	φ^G	Ranking
AI	0.3776	0.4632	0.5026	0.4478	1
Big Data	0.4129	0.4671	0.2207	0.3669	2
CC	0.1158	0.4127	0.3844	0.3043	3
IoT	0.3436	0.3473	0.2087	0.2999	4
EC	0.0214	0.1057	0.2161	0.1144	5
Blockchain	0.2921	−0.0336	0.0594	0.1060	6
CommunS	0.1375	0.0455	0,0985	0.0938	7
Sensor	0.1476	−0.0619	−0.0625	0.0077	8

(*continued*)

Table 7. (*continued*)

Alternatives	DM1 φ	DM2 φ	DM3 φ	$\varnothing^G$	Ranking
CyberS	−0.3287	−0.2184	−0,0431	−0.1967	9
CPS	−0.3143	−0,1226	−0.2072	−0.2147	10
ARS	−0.2630	−0.2884	−0.2099	−0.2538	11
Digital Twins – DT	−0.2795	−0.3646	−0.2064	−0.2835	12
3D Printing	−0.1706	−0.3474	−0.5441	−0,3540	13
IWN	−0.4926	−0.4046	−0.4173	−0.4382	14

5 Discussion and Conclusions

This paper set out to address energy management policies in Brazilian industries. To this end, a multi-criteria group decision problem was structured based on PROMETHEE-ROC for group decision contexts. Thus, three experts from different areas of expertise actively participated in the process as DMs, so that the final recommendation was generated from the concept of global net flow. This element takes into account the individual net flows obtained by using PROMETHEE II and eliciting weights via ROC.

The evaluation conducted in this paper is important for minimizing conflicts and aggregating the preferential information of all DMs for the overall solution recommended. Furthermore, the use of ROC weights minimizes the effort DMs have to spend on the evaluation process, since exact values of weights do not have to be provided. Using this methodology can potentially bring greater precision to the results when compared to a voting procedure, for instance, which has only ordinal information. The PROMETHEE-ROC for Group Decision provides cardinal evaluation for each alternative in terms of overall net flows, which gives more informative evaluation than simply the position of each alternative in the ranking.

The results have shown that the three DMs, besides having different backgrounds and perceptions of the problem, have obtained relatively similar results for the top ranked digital technologies. Those top-ranked alternatives in individual rankings were also obviously reflected as top-ranked in the global group ranking: AI, Big Data, Cloud Computing and IoT were the best evaluated digital technologies. Brazilian industries should focus on implementing such technologies so as to improve their energy management, in order to obtain results such as reducing their energy consumption and costs.

Regarding AI implementation, companies should focus on creating and improving systems that can simulate human thought processes, with or without explicit programming. Big data involves sets of collection, storage, and analysis of complex and varied information in large volumes. Cloud computing consists of hardware, software, and data resources that are stored and accessed remotely through the internet. Finally, IoT focuses on interconnecting physical devices, machines, and objects through the internet.

Finally, it is important to highlight that, depending on the situation and on the results obtained, the application of mechanisms to remedy existing conflicts in the group may be necessary, and, similarly, the use of sensitivity analysis to reduce the uncertainty of the results or statistical tests to assess the robustness of the individual recommendations and the overall recommendation of the group.

It is worth mentioning that this work also faced limitations. For instance, the challenges of implementing DTs in Brazilian industries were not considered, due to the different sectors that needed evaluation. However, since some of the criteria include costs, usability, and job generation, it is expected that barriers associated with high implementation costs of DTs will be found, as well as the poor internet access faced by some Brazilian regions. Workforce resistance is anticipated due to the need for extensive training and knowledge related to the potential loss of some operational jobs that can be replaced by DTs.

Further research is recommended to apply the decision model in Brazilian companies. To maintain the group aspect, the application should consider at least three decision-makers from different areas of the company: one from the energy sector, one from operational activities, and one from senior management. Additionally, the different sectors of Brazilian industries should be considered to compare the results.

Furthermore, for future work it is important to assess the impact of changes in the degree of importance of each DM, as well as applying statistical tests such as Kendall Test to assess the correlation between the global ranking and the individual rankings.

Acknowledgments. The authors are most grateful for CNPq and CAPES, for the financial support provided.

Disclosure of Interests. The authors have no competing interests to declare that are relevant to the content of this article.

References

Adenuga, O.T., Mpofu, K., Boitumelo, R.I.: Energy efficiency analysis modelling system for manufacturing in the context of industry 4.0. Procedia CIRP **80**, 735–740 (2019)

Anastasovski, A.: What is needed for transformation of industrial parks into potential positive energy industrial parks? A review. Energy Policy **173**, 113400 (2023). https://doi.org/10.1016/J.ENPOL.2022.113400

Barron, F.H.: Selecting a best multiattribute alternative with partial information about attribute weights. Acta Physiol. (Oxf) **80**(1–3), 91–103 (1992)

Barron, F.H., Barrett, B.E.: Decision quality using ranked attribute weights. Manag. Sci. **42**(11), 1515–1523 (1996)

Brans, J.-P., Vincke, P.: A preference ranking organisation method. Manag. Sci. **31**(6), 647–656 (1985)

Da Silva, L.B.L., Frej, E.A., De Almeida, A.T., Ferreira, R.J.P., Morais, D.C.: A review of partial information in additive multicriteria methods. IMA J. Manag. Math. **34**(1), 1–37 (2023)

De Almeida Filho, A.T., Clemente, T.R.N., Morais, D.C., de Almeida, A.T.: Preference modeling experiments with surrogate weighting procedures for the PROMETHEE method. Eur. J. Oper. Res. **264**(2), 453–461 (2018)

De Almeida, A.T., Cavalcante, C.A.V., Alencar, M.H., Ferreira, R.J.P., de Almeida-Filho, A.T., Garcez, T.V.: Multicriteria and Multiobjective Models for Risk, Reliability and Maintenance Decision Analysis, vol. 231. Springer, Cham (2015)

Energy Research Company. 2023 Statistical Yearbook of electricity (2023a)

Frank, A.G., Dalenogare, L.S., Ayala, N.F.: Industry 4.0 technologies: implementation patterns in manufacturing companies. Int. J. Prod. Econ. **210**, 15–26 (2019)

Hosouli, S., Gaikwad, N., Qamar, S.H., Gomes, J.: Optimizing photovoltaic thermal (PVT) collector selection: a multi-criteria decision-making (MCDM) approach for renewable energy systems. Heliyon **10**, e27605 (2024). https://doi.org/10.1016/j.heliyon.2024.e27605

International Energy Agency. World Energy Outlook 2023 (2023)

Kshanh, I., Tanaka, M.: Comparative analysis of MCDM for energy efficiency projects evaluation towards sustainable industrial energy management: case study of a petrochemical complex. Expert Syst. Appl. **255**, 124692 (2024). https://doi.org/10.1016/j.eswa.2024.124692

Lyu, W., Liu, J.: Artificial Intelligence and emerging digital technologies in the energy sector. Appl. Energy **303**, 117615 (2021)

Mardani, A., et al.: A review of multi-criteria decision-making applications to solve energy management problems: two decades from 1995 to 2015. Renew. Sustain. Energy Rev. **71**, 216–256 (2017). https://doi.org/10.1016/j.rser.2016.12.053

Maretto, L., Faccio, M., Battini, D.: A multi-criteria decision-making model based on fuzzy logic and AHP for the selection of digital technologies. IFAC-PapersOnLine **55**(2), 319–324 (2022)

Mareschal, B., Brans, J.-P., Macharis, C.: The GDSS PROMETHEE procedure: a PROMETHEE-GAIA based procedure for group decision support. J. Decis. Syst. **7** (1998)

Maroufkhani, P., Desouza, K.C., Perrons, R.K., Iranmanesh, M.: Digital transformation in the resource and energy sectors: a systematic review. Resour. Policy **76**, 102622 (2022)

Maretto, D.C., Almeida, A.T., Alencar, L.H., Clemente, T.R.N., Cavalcanti, C.Z.B.: PROMETHEE-ROC model for assessing the readiness of technology for generating energy. Math. Probl. Eng. **2015**, 1–11 (2015)

Morais, D.C., de Almeida, A.T., Alencar, L.H., Clemente, T.R.N., Cavalcanti, C.Z.B.: PROMETHEE-ROC model for assessing the readiness of technology for generating energy. Math. Probl. Eng. **2015**(1), 530615 (2015)

Pessl, E., Sorko, S.R., Mayer, B.: Roadmap industry 4.0 - implementation guideline for enterprises. In: 26th International Association for Management of Technology Conference, IAMOT, pp. 1728–1743 (2020)

Roth, J., Brown, H.A., IV., Jain, R.K.: Harnessing smart meter data for a Multitiered Energy Management Performance Indicators (MEMPI) framework: a facility manager informed approach. Appl. Energy **276**, 115435 (2020)

Richter, B.K., et al.: Industrial energy efficiency assessment and prioritization model: an approach based on multi-criteria method PROMETHEE. Int. J. Sustain. Energy Plan. Manag. **37**, 41–60 (2023). https://doi.org/10.54337/ijsepm.7335

Schumacher, A., Erol, S., Sihn, W.: A maturity model for assessing industry 4.0 readiness and maturity of manufacturing enterprises. Procedia CIRP **52**, 161–166 (2016)

Sivill, L., Manninen, J., Hippinen, I., Ahtila, P.: Success factors of energy management in energy-intensive industries: development priority of energy performance measurement. Int. J. Energy Res. **37**(8), 936–951 (2013)

Stillwell, W.G., Seaver, D.A., Edwards, W.: A comparison of weight approximation techniques in multiattribute utility decision making. Organ. Behav. Hum. Perform. **28**(1), 62–77 (1981)

Tanveer, U., Kremantzis, M.D., Roussinos, N., Ishaq, S., Kyrgiakos, L.S., Vlontzos, G.: A fuzzy TOPSIS model for selecting digital technologies in circular supply chains. Supply Chain Anal. **4**, 100038 (2023). https://doi.org/10.1016/j.sca.2023.100038

United Nations. The Sustainable Development Goals Report (2023)

Vansnick, J.C.: On the problem of weights in multiple criteria decision making (the noncompensatory approach). Eur. J. Oper. Res. **24**(2), 288–294 (1986)

Stability of Surrogate MCDA Weights Under Different Assumptions on Value Distributions

Sebastian Lakmayer[1] and Mats Danielson[1,2]

[1] Department of Computer and Systems Sciences, Stockholm University, P.O. Box 1203, SE-164 25 Kista, Sweden
mats.danielson@su.se
[2] International Institute for Applied Systems Analysis, IIASA, Schlossplatz 1, AT-2361 Laxenburg, Austria

Abstract. This article studies the effects of sampling the alternative values from different distributions when studying the performance of surrogate weight methods in the additive model in multi-criteria decision analysis. The aim is to demonstrate surrogate weights' performance invariance regarding underlying actual distributions. Multiple distributions are characterised and examined through extensive simulation to evaluate their influence on the efficacy of surrogate weight approaches. It was found that employing the presently accepted standard distributions for alternative values led to outcomes from the surrogate weight methods that were remarkably consistent, barring a few notable deviations—suggesting considerable robustness. In contrast, drawing samples from more extreme distributions resulted in greater divergence. Overall, the observed patterns of the surrogate weight approaches align, to a substantial degree, with findings reported in prior research. We conclude that the performance of the surrogate weight methods is generally stable under a wide variety of reasonable alternative value distributions and show a case when the distribution is too skewed.

Keywords: Multi-criteria decision analysis · Criteria weights · Distribution stability · Surrogate Weights · Ordinal Ranking · Value distributions

1 Introduction

Decision-making is a cornerstone of human existence. Each day, individuals and groups alike are confronted with a multitude of decisions. To support decision-making in both individual and collective contexts, a range of structured methodologies has been devised. A prominent and widely recognised framework in this domain is multi-criteria decision analysis (MCDA). Among the various approaches encompassed by MCDA, the most prevalent is multi-attribute utility theory (MAUT)—referred to as multi-attribute value theory (MAVT) in deterministic contexts, where alternatives do not involve uncertain outcomes [1]. Within this framework, the additive model represents the most frequently applied method for ranking alternatives in accordance with the preferences of the decision-makers (DMs) [2]. In the additive model, the resulting (weighted) value is calculated according to Eq. (1), which is also the model this study focuses on.

$$V(a) = \sum_{i=1}^{n} w_i v_i(a). \qquad (1)$$

In (1), $V(a)$ is the sum product of the values of alternative a for the different criteria where w_i represents the weight (preference) of criterion i out of n criteria and v_i represents the value of alternative a under criterion i. Weights can have a similar function when combining the valuations of different stakeholders in a group decision setting. The different stakeholders can either rank the alternatives individually or supply their preferences for a common ranking. The final ranking can then be obtained either by forming a weighted sum of the individual rankings, by voting, or by a superior DM making the final ranking by itself. For the case of a weighted final ranking, surrogate weights have the same role as in individual MCDA. Figure 1 shows a weight tree containing both criteria weights and stakeholder weights. Thus, while the discussion below uses criteria weights as examples, the results apply equally to group decisions.

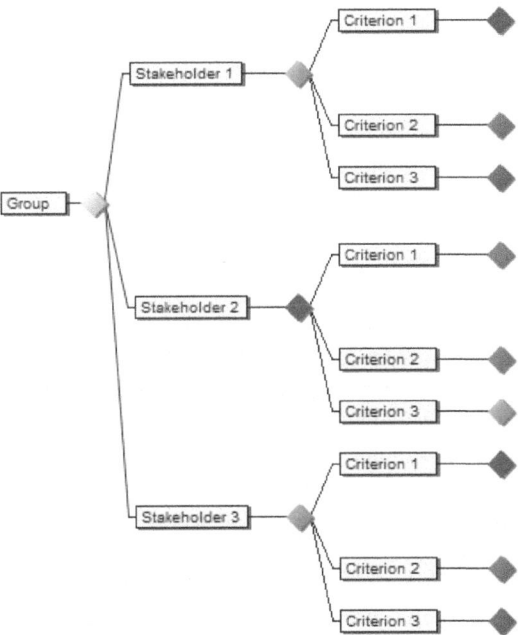

Fig. 1. Weight tree with criteria and stakeholders

The MAVT framework is associated with certain recognised limitations, as noted by several researchers (see, for instance, [3]), notably due to its reliance on input data of a precision that decision-makers (DMs) rarely possess or are able to articulate—thus rendering the resulting choices potentially based on approximate estimates. Various strategies have been proposed to address this issue, including the use of interval-valued or fuzzy representations of data [4], as well as ordinal-level rankings of criteria [1]. The present article focuses on the latter approach, a well-established method for accommodating implicitly expressed preferences.

Thus, in MCDA users are often confronted with imprecise information as input for these methods. MCDA needs information from two different classes of inputs: the weight of each criterion and the value of each alternative with respect to each criterion. For both individual decision-makers and groups of decision-makers, it can be complicated to find or agree on precise numbers for the criteria weights. The reasons behind the inability of a DM to assign precise values to the criteria are manifold [2], especially when many criteria are involved [5–6]. Ranking of the criteria can often be an easier way to express the preferences for the different criteria. The most common method to transform ordinal information into numerical values, which can then be used as criteria weights in the additive model, is the use of surrogate weights. Different surrogate weight methods have been developed over the last decades [7] and the research in this area is still very active [8].

Nonetheless, imprecision in information also arises on the side of alternative values—that is, regarding the actual values alternatives may ultimately assume and the manner in which these are distributed. A method for addressing such uncertainty in alternative values was put forward by [9]. The majority of existing studies have focused primarily on distributions used for generating criteria weights; see, for example, [1, 7, 10]. In contrast, the distributions associated with alternative values have received comparatively limited attention. An earlier contribution by [11] included a comparison of only the normal and uniform distributions in this context. Most other works have not addressed the matter at all. Yet, depending on the characteristics of the decision environment, a variety of alternative value distributions may be plausible or even more suitable. In [12], the present authors employed two such alternative distributions while evaluating the effectiveness of surrogate weights. The results revealed only minor differences in hit ratios, a commonly used metric for assessing the performance of surrogate weight methods.

The primary emphasis of the earlier study lay in comparing the performance of various surrogate weight methods relative to one another, and within that scope, the methods demonstrated a considerable degree of robustness. However, the investigation gave rise to a distinct line of inquiry: Given a set of well-established and reliably performing surrogate weights, to what extent is their performance affected by alterations in the underlying value distributions? That is, the analytical perspective is inverted. Rather than juxtaposing surrogate weight methods, the focus shifts to assessing the collective resilience of surrogate weights when exposed to a variety of value distributions—essentially, examining their stability under distributional diversity. Beyond this conceptual shift, the earlier article explored only two distributions (a U-shaped beta distribution and a lognormal distribution), and observed minor deviations in hit ratios. Naturally, any simulation-based study entails a trade-off between the number of iterations and the breadth of variation introduced. In order to attain a more comprehensive understanding of how alternative value distributions impact surrogate weight performance, the present work reorients the investigation, incorporating a broader range of distributions as well as a greater diversity of alternative and criterion combinations.

Additionally, the weight elicitation can also include variations. Often in the literature, a simple sampling of the weights from a uniform distribution is done with normalisation of the weights, see e.g. [11]. Normalisation or not induces different degrees of freedom (DoF) for the DM. Previous studies show a measurable difference in the performance of surrogate weight methods [12–14], measured by the hit ratio, a standard measure of success [11]. The concept of DoF refers to how a DM arrives at their weights in their internal deliberation processes. If a DM mentally distributes weights from a non-fixed weight pool (requiring a subsequent normalisation), an N DoF process is employed for N criteria. If, on the other hand, a DM mentally uses a fixed-size weight pool to begin with, the last weight is determined by a normalisation constraint, leading to N–1 DoF for N criteria. For a particular DM, it is not known which of the mental models is adhered to, or perhaps a mixture of both. This is why surrogate methods must perform well under both models and for all combinations of those models.

Next, an overview of the employed rank ordering methods, an overview of reasonable distributions for the alternative values, and the weight elicitation under the aspect of different DoF are given. After that, we present the simulation results and conclude the study with remarks and areas for future research.

2 Methodology

The earliest study to systematically introduce simulation as a means of assessing the performance of surrogate weights in depth was [11], although prior contributions such as [15] had already examined the benefits of surrogate weights in considerable detail. In [11], the authors outlined a simulation framework in which both weights and attribute values were generated. However, while the generation of weights was discussed extensively, the description of value generation remained vague, referring merely to "random values [being] the attribute values associated with the [...] alternatives." This limited elaboration is, to some extent, justifiable given that their principal aim was to evaluate the effectiveness of surrogate weight methods. Although the precise mechanism for value generation was not fully specified, the experiments relied on the uniform distribution. As the simulation procedure in [11] subsequently became the standard approach in the literature, the uniform distribution also assumed the status of the default option for value generation. This prevailed whether the researchers used an N–1 DoF weight generator (such as [11]) or an N DoF generator (such as [16], who explicitly state the use of a $U(0,100)$-distribution generator for values).

The next section provides a concise overview of several foundational concepts relevant to the present study. It begins with a presentation of various ordinal ranking techniques. Ordinal ranking constitutes a well-established approach within the additive model of MCDA, particularly suited to situations in which the decision-maker (DM) is either unable or unwilling to specify precise criteria weights. In such contexts, automatically derived surrogate weights can be employed in place of explicitly stated weights for the evaluation of decision alternatives. This article centres on the role of value distributions and their interaction with a representative selection of prominent surrogate weight functions. Accordingly, the surrogate weight methods applied in this analysis include: Rank Sum weights (RS) [17], Rank Order Centroid weights (ROC) [18], and Sum Reciprocal weights (also referred to as Sum Rank, SR) [7].

Second, we describe the concept of weight elicitation from the perspective of different degrees of freedom (DoFs). Third, we present different distributions for the alternative values that seem reasonable within the area of MCDA and the workflow.

2.1 Ordinal Ranking

As noted above, three well-known and well-regarded methods for ordinal ranking are RS [17], ROC [18] and SR [7]. ROC has been shown to be well-performing when the criteria weights are generated under the assumption of N–1 DoF, by which is meant that the simulation of the DM's thought process contains N–1 degrees of freedom when considering N criteria. Thus, it is the DM's mental model that has varying DoF, not the surrogate generation procedures. See [7] for a thorough explanation of the importance of the DoF concept in surrogate weight research. RS, on the other hand, is well-performing when the criteria weights are generated under the assumption of N DoF. The weights for the ROC method are generated according to the formula

$$w_i^{ROC} = 1/N \sum_{j=i}^{N} \frac{1}{j}, i = 1, ..., N; \sum w_i = 1; 0 \leq w_i. \quad (2)$$

The weights for the RS method are generated according to the formula

$$w_i^{RS} = \frac{N+1-i}{\sum_{j=1}^{N}(N+1-j)}, i = 1, ..., N; \sum w_i = 1; 0 \leq w_i. \quad (3)$$

The Sum Reciprocal (SR) method was proposed in [7] in order to alleviate the extreme behaviour of other well-known methods, such as ROC and RS, with respect to different DoF. The weights for the SR method are generated according to the formula

$$w_i^{SR} = \frac{1/i + \frac{N+1-i}{N}}{\sum_{j=1}^{N}\left(1/j + \frac{N+1-j}{N}\right)}, i = 1, ..., N; \sum w_i = 1; 0 \leq w_i. \quad (4)$$

In Eqs. (2), (3), and (4), the parameter N represents the total number of criteria.

2.2 Different Distributions for the Alternative Values

The main focus of this study is to investigate the effect of different distributions of the "true" underlying alternative values on the performance of surrogate weight methods. The use of different distributions for the alternative values was already considered, for example in [11], where both the normal and the uniform distributions were compared, without a significant difference. We found that this result was also mentioned in [21], showing that the consideration of different distributions for alternative values has also been considered in later studies. We showed in [12] that for two different distributions, namely a lognormal and a U-shaped beta distribution, the differences in the hit ratios were relatively small, and the performance of different surrogate weight methods compared to each other was in accordance with previous studies. The hit ratio is calculated by dividing the sum of decision situations where the winning alternative evaluated using the restricted information (ordinal information) is the same as the winning alternative using the true inner weights divided by the number of decision situations. Nevertheless, there exist, in theory and practice, more distributions that are reasonable candidates for the alternative values. Thus, the alternative value distributions studied are briefly discussed next. In order to make the distributions comparable, the values in each criterion are normalised.

At first glance, the uniform distribution $U(0,1)$ may appear to be a reasonable and neutral choice for generating values, as it does not inherently privilege any particular point within its domain and thus seems free of bias. However, upon closer examination, its "naturalness" can be questioned. The assumption underlying $U(0,1)$ is that all values within the specified interval are equally probable—a condition that does not hold in many real-world contexts. For instance, a given value scale may exhibit a central tendency, with mid-range values occurring more frequently, or it may be skewed, with values clustering on one side of the centre. Alternatively, the distribution may exhibit yet another structural pattern. While, in principle, a robust surrogate weight method ought to function independently of the value distribution, the empirical basis for this assumption remains limited, as most existing studies have relied almost exclusively on the uniform distribution. This article seeks to address this gap by examining the stability of several well-established and effective surrogate weight functions when confronted with varying value distributions. The study proceeds as follows: a representative subset of surrogate functions was selected and paired with a set of value distributions exhibiting diverse structural characteristics. The uniform distribution was included as a benchmark. The remaining distributions comprise the normal distribution, several triangular distributions (left-skewed, right-skewed, and symmetric), and the more strongly skewed Erlang distribution. Triangular distributions, in particular, are frequently employed in modelling decision-maker value preferences, as they are intuitively accessible, require no parameter estimation, and are fully defined by three points—making them representative of a wide range of centrally distributed functions [19]. A substantial number of simulations were then conducted across the combinations of weight and value distributions, yielding a series of output datasets. These results are presented in Sect. 4.1, both in tabular form for reference and as graphical representations for ease of interpretation. This is followed by a discussion of the findings and a summary of the principal conclusions.

The Uniform Distribution

This distribution is the default distribution in literature for generating alternative values, see e.g., [10, 13, 14, 20 – 22].

The Normal Distribution

While already [11] used a normal distribution for the alternative values and compared the effect of using it instead of a uniform distribution, it was not a main focus of that article and it did not constitute a larger structured study. We use the normal distribution as a benchmark distribution in this study alongside the uniform distribution.

The Triangular Distribution

The triangular distribution is useful to model cases where the decision-maker can only provide little information about the alternative values. The DM only has to provide a minimum, middle, and maximum value to completely specify the distribution. Then, the alternative values are modelled by a triangular distribution with the middle value as the mode. We use a right-skewed, a symmetric, and a left-skewed triangular distribution in our comparisons. The lower limit is 0, the upper limit is 1 and the mode is then 0.25, 0.5, or 0.75.

The Erlang Distribution
The Erlang distribution is a special case of the Gamma distribution. In order to cover more extreme distributions, we also used an Erlang distribution with parameters $k = 2$ and $\lambda = 1$. This led to a right-skewed distribution, resulting in a tendency towards comparatively large alternative values. The distribution has been popular in PERT applications, among others. The probability density function of the Erlang distribution is given by Eq. 5 as

$$f(x; k, \lambda) = \frac{\lambda^k x^{k-1} e^{-\lambda x}}{(k-1)!} \text{ for } x, \lambda \geq 0. \tag{5}$$

A Mixed Distribution
Finally, we also combined all discussed distributions into a compound one for reference. For each criterion, one of the above-mentioned distributions is chosen randomly with equal probability in each iteration with replacement. Then the alternative values for each criterion are sampled from the selected distribution. Thus, the distribution is kept the same within one simulation loop but varies across the loops.

3 Discussion

This study investigated the impact of varying alternative value distributions on the performance of surrogate weight models within the additive MCDA framework. The analysis was conducted by systematically exploring multiple configurations involving the number of alternatives, the number of criteria, degrees of freedom, and differing forms of value distributions. While two such distributions had previously been examined by the authors in [12], the present work builds upon and substantiates that earlier investigation by incorporating a broader spectrum of value distributions representative of plausible decision-making scenarios, along with the inclusion of the more pronouncedly skewed Erlang distribution. Furthermore, this study examined the variability in outcomes resulting from repeated simulations using distinct random seeds. To the best of the authors' knowledge, such a comparative analysis has not been previously undertaken. Based on the application of realistic alternative value distributions presented herein, future developers of decision-making models may place greater confidence in the robustness of surrogate weight methods across a diverse array of distributional conditions.

4 Results

4.1 Comparison of Different Distributions for the Alternative Values

In this article, we compare several combinations of alternatives and criteria for different distributions of alternative values, as well as a mix of distributions for the alternative values. Altogether, nine different combinations of number of alternatives and number of (criteria) weights were tested: 5 (alternatives)/3 (weights), 5/15, 10/6, 10/12, 15/9, 20/6, 20/12, 25/3, and 25/15, of which five are presented as Tables 1, 2, 3, 4 and 5 and Graphs 1 – 5. The excluded combinations are similar in performance to the included ones and do not contribute significantly to the overall conclusions of the study. The results for each combination are grouped according to the different DoFs of the generators used: N DoF, N–1 DoF and an equal mix of both. The concept of generators is described in detail in [12]. In order to measure variation in the simulations, each of these combinations was modelled in ten batches with 100,000 decision situations (combinations of weight vector and alternative values matrices) in each, resulting in 1 million simulations for each aforementioned combination. In the tables and graphs below, "Mixed DoF" means that the generator was sampled for each criteria (or stakeholder) weight simulation with a probability of 0.5 from either the N-generator or the N–1-generator.

For each configuration of alternatives and criteria, a table is initially presented that quantifies the deviation of each alternative value distribution, as well as the aggregated distribution set, from the overall mean hit ratio associated with the uniform distribution for the corresponding configuration. Specifically, the mean ratio of the uniform distribution—serving as the historical benchmark in this study—is subtracted from the mean ratio of each respective distribution. Subsequently, a graph is provided for each examined combination of alternatives and (criteria) weights, in which the hit ratios from individual simulation batches are displayed as discrete data points.

Table 1 and Graph 1. Differences in hit ratios for 5 alternatives and 15 criteria (5/15).

DoF	Method	mean_uniform	diff_erlang	diff_normal	diff_tri25	diff_tri50	diff_tri75	diff_mix
N-1 DoF	ROC	87.76	-0.46	-0.09	-0.30	-0.08	0.19	-0.12
	RS	78.26	-0.90	-0.21	-0.47	-0.09	0.40	-0.21
	SR	85.32	-0.32	-0.12	-0.24	-0.10	0.10	-0.11
Mixed DoF	ROC	83.14	-0.73	-0.04	-0.39	-0.08	0.29	-0.21
	RS	85.07	-0.42	-0.06	-0.24	-0.07	0.21	-0.21
	SR	85.56	-0.54	-0.14	-0.26	-0.04	0.20	-0.19
N DoF	ROC	78.60	-1.18	-0.19	-0.49	-0.10	0.45	-0.38
	RS	91.84	-0.08	0.04	0.02	0.03	0.03	-0.07
	SR	85.78	-0.87	-0.07	-0.35	-0.06	0.35	-0.30

Table 2 and Graph 2. Differences in hit ratios for 10 alternatives and 6 criteria (10/6).

DoF	Method	mean_uniform	diff_erlang	diff_normal	diff_tri25	diff_tri50	diff_tri75	diff_mix
	ROC	82.18	-0.54	-0.60	-0.48	-0.49	-0.22	-0.38
N-1 DoF	RS	76.82	-1.88	-1.16	-1.10	-0.96	-0.37	-0.83
	SR	80.05	-0.17	-0.57	-0.28	-0.48	-0.43	-0.28
	ROC	79.68	-2.30	-1.00	-1.06	-0.80	-0.07	-0.81
Mixed DoF	RS	80.79	-1.31	-0.68	-0.61	-0.53	-0.16	-0.54
	SR	81.27	-1.24	-0.61	-0.58	-0.50	-0.17	-0.47
	ROC	77.44	-3.52	-1.58	-1.78	-1.14	0.19	-1.45
N DoF	RS	85.08	-0.43	-0.38	-0.25	-0.18	0.03	-0.31
	SR	82.75	-1.85	-0.85	-0.94	-0.54	0.18	-0.74

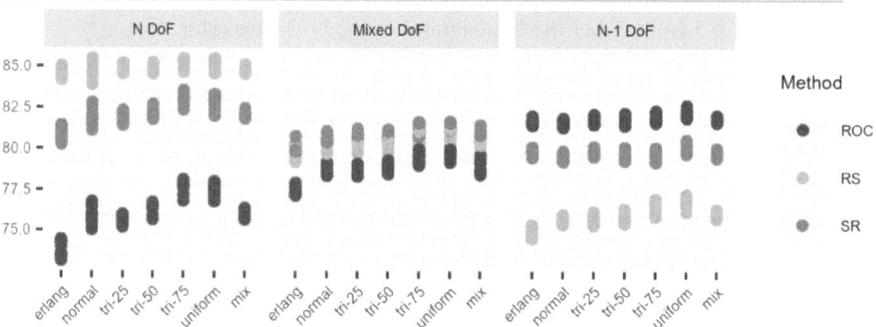

It is evident that the variation across simulation batches employing identical parameters and surrogate methods is markedly lower than the variation observed between different surrogate methods or levels of degrees of freedom. This aligns with expectations and confirms that the experimental outcomes exhibit stability with respect to the number of simulations conducted—that is, the study's scale is methodologically adequate. Were the value distributions to exert no influence at all on the decision outcomes produced by surrogate methods, then all data points of the same colour within each graph would align along a straight line, aside from minor fluctuations attributable to simulation variability. Upon examination of the plots, this pattern is largely observed, with deviations remaining within the expected bounds of stochastic variation. The notable exceptions are ROC, which for N DoF in particular is dependent on the underlying value distribution, and to a lesser extent RS for N–1 DoF. This once more underlines that the design desiderata of SR – to be as stable as possible under as many different conditions as possible – also hold for varying value distributions.

With regard to the surrogate weight methods themselves, it is clearly observable that in non-mixed degrees of freedom scenarios, the variation across methods exceeds the variation observed across batches for each individual distribution or distribution mix. Although some overlap occurs among the batches corresponding to different surrogate

Table 3 and Graph 3. Differences in hit ratios for 10 alternatives and 12 criteria (10/12).

DoF	Method	mean_uniform	diff_erlang	diff_normal	diff_tri25	diff_tri50	diff_tri75	diff_mix
N-1 DoF	ROC	84.06	-1.21	-0.56	-0.69	-0.47	0.07	-0.44
	RS	74.17	-2.21	-0.89	-1.17	-0.68	0.35	-0.72
	SR	81.29	-0.80	-0.52	-0.49	-0.44	-0.13	-0.35
Mixed DoF	ROC	79.30	-2.07	-0.73	-1.05	-0.50	0.42	-0.64
	RS	81.27	-1.27	-0.50	-0.62	-0.32	0.18	-0.38
	SR	81.88	-1.43	-0.51	-0.73	-0.35	0.26	-0.44
N DoF	ROC	74.39	-2.65	-0.69	-1.32	-0.55	0.76	-0.78
	RS	88.26	-0.03	-0.00	-0.02	0.05	0.02	-0.01
	SR	82.38	-1.89	-0.38	-0.95	-0.30	0.58	-0.44

methods, the overall performance patterns align with findings reported in previous literature from the perspective of weight construction. No single distribution consistently outperforms the others across all examined configurations. Among the methods considered, the Sum Reciprocal (SR) approach emerges as the most adaptable, exhibiting neither pronounced performance declines nor exceptional volatility, and demonstrating the highest average effectiveness across the various settings.

Although the reasoning presented thus far has been primarily qualitative, equivalent conclusions can be reached through a more formal quantitative analysis. The standard deviations calculated within sets of simulations sharing identical parameters are substantially lower than those observed between sets with differing parameters. This aligns with the visual proximity of data points associated with the same parameter configurations. Furthermore, the standard deviations across different value distributions—under fixed conditions of degrees of freedom, alternatives, and (criteria) weights—are also significantly smaller than those observed when one or more of these parameters vary. This supports the visual impression that the plotted data, with the possible exception of the Erlang distribution, conform closely to a linear trend within each plot.

Table 4 and Graph 4. Differences in hit ratios for 15 alternatives and 9 criteria (15/9).

DoF	Method	mean_uniform	diff_erlang	diff_normal	diff_tri25	diff_tri50	diff_tri75	diff_mix
	ROC	81.60	-1.37	-0.96	-0.86	-0.76	-0.21	-0.67
N-1 DoF	RS	73.24	-2.89	-1.72	-1.57	-1.29	-0.34	-1.18
	SR	78.79	-0.68	-0.76	-0.47	-0.62	-0.44	-0.39
	ROC	77.62	-2.78	-1.30	-1.45	-0.98	0.03	-0.96
Mixed DoF	RS	79.36	-1.54	-0.87	-0.86	-0.65	-0.09	-0.63
	SR	79.94	-1.65	-0.86	-0.89	-0.63	-0.06	-0.58
	ROC	73.74	-4.28	-1.81	-2.14	-1.40	0.19	-1.62
N DoF	RS	85.41	-0.17	-0.18	-0.12	-0.10	-0.01	-0.15
	SR	81.03	-2.62	-1.01	-1.33	-0.72	0.26	-0.98

5 Concluding Remarks

The simulation results indicate that, for the majority of alternative value distributions, the variability in hit ratios remains comparable across identical parameter configurations—that is, consistent combinations of the number of alternatives, number of weighs (criteria or stakeholder), and degrees of freedom. In general, the relative performance of the surrogate weight methods remains highly consistent across the scenarios examined. In settings involving mixed degrees of freedom, the Sum Reciprocal (SR) method typically exhibits the highest level of performance. At the very endpoints of the DoF spectrum, where more extreme weights are generated, as in the case of N–1 DoF, the ROC method generally performs the best and for N DoF, the RS method generally performs the best. The results of this study indicate that surrogate weights can be used in combination with the most common distributions of alternative values without a lot of variation in the performance, measured by the hit ratio, compared to each other. Again, as already noted in [12], the performance difference between N and N–1 DoF cannot be eliminated and the actual DoF for a particular decision situation cannot be known. As observed, most surrogate weights cater more to one DM mental model (DoF) as shown by the tables and plots. Only if explicitly designed to handle all DoF combinations well, surrogate methods (such as SR) will be robust over differing DoFs. To cater for the DoF uncertainty inherent in all decision situations, and likewise for the uncertainty regarding the DM's actual distribution of values along the value scales, the SR method is the most stable. However, should another surrogate method be desired, this study shows that – within reasonable

Table 5 and Graph 5. Differences in hit ratios for 20 alternatives and 12 criteria (20/12).

DoF	Method	mean_uniform	diff_erlang	diff_normal	diff_tri25	diff_tri50	diff_tri75	diff_mix
	ROC	81.47	-1.78	-1.00	-0.99	-0.75	-0.19	-0.67
N-1 DoF	RS	70.54	-3.58	-1.74	-1.81	-1.35	-0.15	-1.29
	SR	78.03	-1.04	-0.77	-0.58	-0.64	-0.36	-0.46
	ROC	76.19	-3.17	-1.44	-1.68	-1.16	0.09	-1.07
Mixed DoF	RS	78.23	-1.78	-0.85	-0.97	-0.75	-0.08	-0.62
	SR	78.93	-2.11	-0.95	-1.20	-0.80	-0.00	-0.80
	ROC	71.09	-4.72	-1.97	-2.21	-1.39	0.30	-1.69
N DoF	RS	86.10	-0.20	-0.08	-0.11	-0.06	0.03	-0.12
	SR	79.95	-3.44	-1.21	-1.56	-0.80	0.44	-1.13

limits – value distributions is a factor that many surrogate weight approaches are able to deal with well. If the reader wants to try out these and other surrogate weights, an open-source software library (UNEDA) is bundled together with [19]. Altogether, in this article we showed that for reasonable distributions for the alternative values the impact on the performance of the surrogate weight methods is small. Generally, the choice of the specific surrogate weight method and the DoF have a larger impact. However, only a selected range of combinations of the number of (criteria/stakeholder) weights and alternatives has been tested. Further, the choice of distributions for the alternative values was limited. Hence, these limitations should be kept in mind when considering the outcomes of this study.

6 Further Research

Although surrogate weight methods have been the subject of scholarly inquiry for a considerable period, the field remains one of ongoing and active development. Numerous facets have yet to be examined in depth. While the present study concentrates on varying alternative value distributions, it does not address the potential effects of filtering extreme weights, as proposed in [12]. Moreover, this investigation is confined to ordinal ranking approaches; the implications for preference strength methods—also referred to as cardinal ranking—remain unexplored. In addition, the analysis has relied exclusively on the widely adopted hit ratio as the measure of performance. A range of alternative performance metrics exists—see, for instance, [8], which provides a comprehensive overview of methods and associated quality measures. Broadening the scope of this research to incorporate additional performance indicators constitutes a promising direction for future work.

Acknowledgements. The author SL would like to thank the late Professor Love Ekenberg of Stockholm University for introducing SL to the area of surrogate weights in MCDA at the time when LE was the supervisor of SL's thesis work. The author MD was partially funded by the European Commission research programme Horizon Europe, grant agreement 101074075.

Disclosure of Interests. The authors have no competing interests to declare relevant to this article's content.

References

1. Danielson, M., Ekenberg, L.: A robustness study of state-of-the-art surrogate weights for MCDM. Group Decis. Negot. **26**, 677–691 (2016)
2. Aguayo, E.A., Mateos, A., Jiménez, A.: A new dominance intensity method to deal with ordinal information about a DM's preferences within MAVT. Knowl.-Based Syst. **69**, 159–169 (2014)
3. Von Winterfeldt, D., Edwards, W.: Decision Analysis and Behavioral Research. Cambridge University Press, Cambridge (1986)
4. Danielson, M.: Computational Decision Analysis. Ph.D. Thesis, Royal Institute of Technology, Stockholm, Sweden (1997)
5. Park, K.S.: Mathematical programming models for characterizing dominance and potential optimality when multicriteria alternative values and weights are simultaneously incomplete. IEEE Trans. Syst. Man Cybern. Part A: Syst. Hum. **34**(5), 601–614 (2004)
6. Larsson, A., Riabacke, M., Danielson, M., Ekenberg, L.: Cardinal and Rank ordering of criteria – addressing prescription within weight elicitation. Int. J. Inf. Technol. Decis. Mak. **14**(6), 1299–1330 (2014)
7. Danielson, M., Ekenberg, L.: Rank ordering methods for multicriteria decisions. In: Zaraté, P., Kersten, G.E., Hernández, J.E. (eds.) Group Decision and Negotiation. A Process-Oriented View. GDN 2014. Lecture Notes in Business Information Processing, vol. 180, pp. 128–135. Springer, Cham (2014)
8. Chergui, Z., Jiménez-Martín, A.: On ordinal information-based weighting methods and comparison analyses. Information **15**, 527 (2024)
9. Sarabando, P., Dias, L.: Simple procedures of choice in multicriteria problems without precise information about the alternatives' values. Comput. Oper. Res. **37**, 2239–2247 (2010)
10. Danielson, M., Ekenberg, L.: The CAR method for using preference strength in multicriteria decision making. Group Decis. Negot. **25**, 775–797 (2016)
11. Barron, F., Barrett, B.: Decision quality using ranked attribute weights. Manag. Sci. **42**(11), 1515–1523 (1996)
12. Lakmayer, S., Danielson, M., Ekenberg, L.: Upper performance limits and distribution invariance for surrogate weights in MCDA. In: Campos Ferreira, M., Wachowicz, T., Zaraté, P., Maemura, Y. (eds.) Human-Centric Decision and Negotiation Support for Societal Transitions, pp. 89–101. Springer, Cham (2024)
13. Lakmayer, S., Danielson, M., Ekenberg, L.: Aspects of ranking algorithms in multicriteria decision support systems. In: Fujita, H., Guizzi, G. (eds.) New Trends in Software Methodologies, Tools and Techniques, pp. 63–75. IOS Press, Amsterdam (2023)

14. Lakmayer, S., Danielson, M., Ekenberg, L.: Automatically generated weight methods for human and machine decision-making. In: Fujita, H., Wang, Y., Xiao, Y., Moonis, A. (eds.) Advances and Trends in Artificial Intelligence. Theory and Applications. IEA/AIE 2023. Lecture Notes in Computer Science, vol. 13925, pp. 195–206. Springer, Cham (2023)
15. Edwards, W., Barron, H.F.: SMARTS and SMARTER: improved simple methods for multiattribute utility measurement. Organ. Behav. Hum. Decis. Process. **60**, 306–325 (1994)
16. Roberts, R., Goodwin, P.: Weight approximations in multi-attribute decision models. J. Multi-Crit. Decis. Anal. **11**, 291–303 (2002)
17. Stillwell, W., Seaver, D., Edwards, W.: A comparison of weight approximation techniques in multiattribute utility decision making. Organ. Behav. Hum. Perform. **28**(1), 62–77 (1981)
18. Barron, F.H.: Selecting a best multiattribute alternative with partial information about attribute weights. Acta Physiol. (Oxf) **80**(1–3), 91–103 (1992)
19. Danielson, M.: Foundations of Computational Decision Analysis, 2nd ed. Sine Metu, Stockholm (2025)
20. Ahn, B.S., Park, K.S.: Comparing methods for multiattribute decision making with ordinal weights. Comput. Oper. Res. **35**(5), 1660–1670 (2008)
21. Sarabando, P., Dias, L.: Multi-attribute choice with ordinal information: a comparison of different decision rules. IEEE Trans. Syst. Man Cybern. Part A: Syst. Hum. **39**, 545–554 (2009)
22. Salo, A.A., Hämäläinen, R.P.: Preference ratios in multiattribute evaluation (PRIME)-elicitation and decision procedures under incomplete information. IEEE Trans. Syst. Man Cybern. Part A: Syst. Hum. **31**(6), 533–545 (2001)

The Approbatory Social Welfare Function: First Results

Gustavo Santos-García and José Carlos R. Alcantud(✉)

BORDA Research Group and IME, University of Salamanca, 37007 Salamanca, Spain
{santos,jcr}@usal.es

Abstract. Since 2009 when preference-approval structures were introduced in the social choice literature, they have been studied and applied to produce explicit voting rules, measure consensus (which in turn requires the construction of appropriate distances or metrics), or clustering, among other areas. This contribution builds on the notion of consistent preference-approval structures. We discuss their relationship to other literature and argue that their introduction is justified because they are at the root of a smooth extension of the Arrovian approach that acts on preference-approval structures, namely, approbatory social welfare functions. Also, our contribution lays out a general two-step procedure that generates approbatory social welfare functions. We prove that this parameterized process enables us to produce Borda-type extensions in this setting, inclusive of existing proposals for the aggregation of complete preorders. Fully developed examples are given, along with incompatible properties that bound the scope of utilization of the novel approbatory social welfare functions.

Keywords: Ranking · Approval · Social welfare function · Voting · Dictatorship

JEL Classification: D63 · D71

1 Introduction

Problems with linearly ranking alternatives in a consistent manner have led to multiple choice designs that consider generalizations and variations of this model. In voting rules, the strengths of approval voting [9] and dis&approval voting [5], best-worst methods [12], and preference-approval structures [10] were used in many studies. The latter model is the subject of our investigation. It relaxes linearity by allowing ties, and simultaneously presents a subset of approved alternatives that is compatible with the order. Following Brams and Sanver [10], other authors studied preference-approval structures. The measurement of consensus was the topic of Erdamar et al. [11]. Albano et al. [1,2] defined distances and proposed clustering strategies. Santos-García and Alcantud [22] proved an exact formula for the number of preference-approval structures (as a function of the numbers of total and approved alternatives), as well as an extension of Arrow's characterization of dictatorship.

Preference-approval structures were introduced with the purpose of defining new voting systems. Our work is motivated by this achievement and also by considerations of bounded rationality. We recall that in a multi-self interpretation of choice models [3,18], *rationales* may capture different 'selves' or 'frames of mind'. Again, these multiple selves can be expressed by either linear orders or other preference models. Final decisions stem from aggregation of these rationales, quite similar to the case of voting systems [3]. However, little has been said about the role of preference-approval structures in this framework, with exceptions such as the seminal article by Brams and Sanver [10] as well as Santos-García and Alcantud [22]. To fill this gap, and considering the adapted version of the Arrovian aggregation framework, we work with approbatory social welfare functions (SWFs). We propose a structured methodology that produces a family of approbatory SWFs. Finally, we prove that some further incompatibilities appear with properties that are natural in popular SWFs.

2 Notation and Preliminaries

Henceforth $X = \{x_1, \ldots, x_n\}$ denotes a set of alternatives. $\mathbf{W}(X)$ denotes the set of all complete preorders (reflexive, complete, transitive) on X. The strict part of $R \in \mathbf{W}(X)$ is denoted by P, and its indifference is I. A tuple $\mathcal{R} \in \mathbf{W}(X)^m$ for some m is a profile of complete preorders. $\mathcal{W}(X) = \bigcup_{m>0} \mathbf{W}(X)^m$ denotes the set of all profiles of complete preorders. Each $\mathcal{R} = (R_1, \ldots, R_m) \in \mathcal{W}(X)$ and $A \subseteq X$ define the restriction of $\mathcal{R}$ to A, and we denote it as $\mathcal{R}|_A = ((R_1)|_A, \ldots, (R_m)|_A)$.

We are concerned with preference-approval structures on X. By this we mean (R, A), with R being a complete preorder on X, and $A \subseteq X$ being the "approved" alternatives in X. Both elements are related by the property that xRy and $y \in A$ oblige to $x \in A$, for all $x, y \in X$. $\mathbf{W}_A(X)$ denotes the set of all preference-approval structures on X with approved alternatives A. Semantically, R captures an "as good as" relation, therefore in preference-approval structures, alternatives that are weakly preferred to an "approved" alternative must be "approved" as well. Henceforth we refer to $\mathbf{W}^A(X) = \{R \in \mathbf{W}(X) \mid (R, A) \in \mathbf{W}_A(X)\}$, the set of complete preorders that generate preference-approval structures on X with approved alternatives A.

Remark 1. Although the term "approved" is firmly established, it is only a convenient default. It can be used to tell apart for example, local and foreign products.

2.1 The Consistent Preference-Approval Model

To extend preference-approval structures to multi-choice situations, we introduce the following related concept:

Definition 1. *A consistent preference-approval profile on X is $(\mathcal{R}, A)$ where $\mathcal{R} = (R_1, \ldots, R_m)$ and each (R_j, A) is a preference-approval structure on X. The set of all consistent preference-approval profiles on X with approved alternatives A is $\mathcal{W}_A(X) = \bigcup_{m>0} \mathcal{W}_A^m(X)$, where $\mathcal{W}_A^m(X) = \mathbf{W}_A(X)^m$ denotes the consistent preference-approval profiles on X with approved alternatives A formed by m preferences.*

Semantic Interpretation. In a social context, these profiles exhibit a minimum level of consistency across agents: a set of "approved" alternatives is common knowledge (e.g., 'local' vs. 'foreign' products), and individual preferences are consistent with this selection. In a multi-self framework of bounded rationality, the decision-maker has definitely distinguished approved from non-approved alternatives (e.g., 'familiar' vs. 'alien' topics in a conference), but she is unsure of the order in which the alternatives are ranked. Nevertheless, the rankings are always consistent with approval.

Henceforth we write $\mathcal{W}^A(X) = \{\mathcal{R} \in \mathcal{W}(X) \mid (\mathcal{R}, A) \in \mathcal{W}_A(X)\}$, the set of profiles of complete preorders that generate consistent preference-approval profiles on X with approved alternatives A.

2.2 Relationship with the Literature

When Brams and Sanver [10] introduced preference-approval structures, their goal was to produce two new voting methods with a hybrid input, namely, preference approval voting and fallback voting. Our research draws a bridge between preference-approval structures and the stream of literature that restricts the domain of preference profiles in order to escape the negative conclusions of Arrow's characterization of dictatorships [6]. Indeed, Arrovian SWFs defined on the unrestricted set of preference profiles that satisfy two desirable properties must be dictatorial. Arrow himself proposed single-peakedness as a restriction that guarantees the transitivity of the simple majority rule, which fails to hold true with the universal set of preference profiles. Structured preferences that serve similar purposes include the suggestion by Inada [17, Section 4], who defined when a complete preorder is separable into two subsets of X. Any such preference can be easily identified with a preference-approval structure on X. A list of these structured preferences is a group-separable preference *profile*. Ballester and Haeringer [7, Theorem 2] proved characterization of group-separable profiles *of linear orders*. It is worthy of note that Inada only used group-separable preference profiles to study the simple majority rule [17, Theorem 4]. And in this collective setting, they do not necessarily deem all elements in one of the subsets preferred to the elements in its complement, for all individual preferences. Therefore, Inada's possibility theorem for the simple majority rule does not need *consistent* preference-approval profiles to hold true. This concept remained undefined until Santos-García and Alcantud [22] first considered it.

A less related subject is ranked soft sets. This soft computing model was proposed by the authors in [21] to model slightly structured situations for multi-criteria decisions. There are two differences that should be emphasised in relation to consistent preference-approval profiles. First, ranked soft sets can be identified with a type of group-separable preference profiles, which are not necessarily consistent preference-approval profiles. Secondly, the complete preorders representing the criteria declare all "non-approved" options indifferent.

To conclude, we mention the role of combinatorics in related literature. Note that there is a natural bijection from $\mathbf{W}^A(X)$ to $\mathbf{W}_A(X)$, for fixed $A \subseteq X$. The number of elements in these sets is computed by the authors in [22].[1] Relatedly, Karpov [19] counted the number of group-separable profiles of *linear orders*. This concept is more general than linear orders separable into two subsets, but it does not allow for ties as in [22], and it does not correspond to consistent preference-approval profiles, as explained above.

3 Approbatory Social Welfare Functions

In this section, we are concerned with the following concept that generalizes Arrovian social welfare functions to consistent preference-approvals:

Definition 2. *[22] An A-approbatory social welfare function (A-ASWF for short) is a mapping $F : \mathcal{W}_A(X) \longrightarrow \mathbf{W}_A(X)$.*

First, we investigate the fundamental properties of this notion. Then we describe explicit procedures for the construction of A-ASWFs. Finally we present simple incompatibilities for mappings as in Definition 2.

3.1 Properties of the A-Approbatory Social Welfare Functions

Let us fix $F : \mathcal{W}_A(X) \longrightarrow \mathbf{W}_A(X)$, an A-ASWF.

We write $F(\mathcal{R}, A) = (F'(\mathcal{R}), A)$ for each $(\mathcal{R}, A) \in \mathcal{W}_A(X)$. Then F' naturally defines a standard SWF on the restricted domain of profiles of complete preorders $\mathcal{R}$ such that $(\mathcal{R}, A)$ is a consistent preference-approval profile, i.e., $F' : \mathcal{W}^A(X) \longrightarrow \mathbf{W}^A(X)$. Using this tool, we say that F satisfies:

1. Unanimity when F' satisfies unanimity: if $R = F'(\mathcal{R})$ for $\mathcal{R} = (R_1, \ldots, R_m) \in \mathcal{W}^A(X)$, and $x_i, x_j \in X$ are such that for any individual order $R_k \in \mathbf{W}^A(X)$ it is the case that $x_i P_k x_j$, then $x_i P x_j$.
2. Independence of Irrelevant Alternatives (IIA) when F' satisfies IIA: if $x_i, x_j \in X$ are such that the profiles $\mathcal{R} \in \mathcal{W}^A(X)$ and $\mathcal{R}' \in \mathcal{W}^A(X)$ rank x_i, x_j the same,[2] it is the case that $F'(\mathcal{R})$ and $F'(\mathcal{R}')$ rank x_i, x_j the same too.

---

[1] At the end of this chapter, the Data Availability statement includes a link to a Mathematica notebook with the calculations producing the exact number of preference-approval structures.
[2] By this we mean that if we write $\mathcal{R} = (R_1, \ldots, R_m)$ and $\mathcal{R}' = (R'_1, \ldots, R'_m)$, then $x_i R_k x_j$ if and only if $x_i R'_k x_j$, for each $k = 1, \ldots, m$.

3.2 Producing A-Approbatory Social Welfare Functions

Basic examples of A-ASWFs include the following shortlist:

1. Constant A-ASWFs: for a fixed preference approval structure (R, A), the map assigning each $(\mathcal{R}, A) \in \mathcal{W}_A(X)$ with (R, A) is an A-ASWF.
2. Dictatorial A-ASWFs: they are defined by a mapping $d : \mathbb{N} \longrightarrow \mathbb{N}$ such that $d(m) \leqslant m$ defines the identity of the dictator when there are m complete preorders, for each m. Any such d defines $F(\mathcal{R}, A) = (R_{d(m)}, A)$ when $(\mathcal{R}, A) \in \mathcal{W}_A^m(X)$ and $\mathcal{R} = (R_1, \ldots, R_m)$.
3. Pairwise dictatorial A-ASWFs: they are defined by $c : \mathbb{N} \longrightarrow \mathbb{N} \times \mathbb{N}$ such that $c(m) = (c_1(m), c_2(m)) \leqslant (m, m)$ defines the identity of a pair of dictators when there are m complete preorders.
   Any such c defines $F(\mathcal{R}, A) = (R_{c(m)}^{\mathcal{R}}, A)$ when $\mathcal{R} = (R_1, \ldots, R_m)$. $R_{c(m)}^{\mathcal{R}}$ is the complete preorder such that: for $x_i, x_j \in X$,

$$x_i \, R_{c(m)}^{\mathcal{R}} \, x_j \Leftrightarrow \begin{cases} x_i \in A, \, x_j \notin A, & \text{or} \\ x_i, x_j \in A, \, x_i \, R_{c_1(m)} \, x_j, & \text{or} \\ x_i, x_j \notin A, \, x_i \, R_{c_2(m)} \, x_j. \end{cases} \quad (1)$$

Recall that Arrow established that in the case of universal domain of complete preorders, unanimity and IIA lead to dictatorial SWFs. Likewise, Theorem 1 below recalls that pairwise dictatorial A-ASWFs are the only A-approbatory SWFs that satisfy unanimity and IIA [22]. Thus we can say that consistent preference-approval profiles with non-empty, non-total sets of approved alternatives, define a domain of preferences permitting to narrowly escape Arrow's incompatibility.

We proceed to define more structured constructions leading to A-approbatory social welfare functions.

In their extensive discussion about the assignment of positions to elements ordered by complete preorders, García-Lapresta and Martínez-Panero [13] considered the following concept:

Definition 3. *Let $M(X)$ denote the set of mappings from X to $\mathbb{R}$. A position operator is a function $\mathcal{O} \colon \mathbf{W}(X) \longrightarrow M(X)$. For simplicity, we denote $\mathcal{O}(R) = O_R \colon X \longrightarrow \mathbb{R}$ whenever $R \in \mathbf{W}(X)$.*

The position operator $\mathcal{O}$ satisfies monotonicity if for all $R \in \mathbf{W}(X)$ and $x_i, x_j \in X$, $x_i R x_j$ if and only if $O_R(x_j) \geqslant O_R(x_i)$ [13, Definition 6]. A weaker property is equality, which amounts to: for all $R \in \mathbf{W}(X)$ and $x_i, x_j \in X$, $x_i I x_j$ implies $O_R(x_j) = O_R(x_i)$.

Remark 2. Conversely, each mapping $U \colon X \longrightarrow \mathbb{R}$ can be regarded as the utility function of the complete preorder defined by $x_i R x_j$ if and only if $U(x_i) \geqslant U(x_j)$, for each $x_i, x_j \in X$. Or dually, it may be considered as the evaluation of a monotonic position operator for the complete preorder R_U defined by

$$x_i \, R_U \, x_j \text{ if and only if } U(x_i) \leqslant U(x_j), \text{ for each } x_i, x_j \in X. \quad (2)$$

Example 1. Examples of the action of position operators on a complete preorder R include the next cases:

1. Kendall [20] considered both the *standard rank* and the *fractional rank*, also called mid-rank.
   In the standard rank, each alternative occupies the highest position in linearly-ordered tie-breaking options. Concerning its utilization, Alcantud et al. [4] and González-Arteaga et al. [15] resorted to this rank in their analysis of the measurement of consensus.
   In the fractional rank, each alternative is given the average position that occupies considering all linearly-ordered tie-breaking options. Gärdenfors [14, Definition 3.3] used a related notion to define the (standard) Borda function intended for complete preorders.[3]
2. In the *modified rank*, each alternative occupies the lowest position in linearly-ordered tie-breaking options. Gärdenfors [14, Definition 7.1] used a related notion to define the restricted Borda function intended for complete preorders [13].
3. García-Lapresta and Martínez-Panero [13] proved two characterizations of the *dense rank*. To define it, let $p_i = |\{x_j \in X \mid x_i P x_j\}|$ denote the number of alternatives strictly dominated by x_i. Then $T_p = \{x_i \in X \mid p_i = p\}$ captures the alternatives that are strictly preferred over exactly p alternatives. When $T = \{p \in \{0, 1, \ldots, n-1\} \mid T_p \neq \emptyset\}$, the dense rank is defined by

$$\forall x_i \in X, O_R(x_i) = |\{\bar{p} \in T \mid p < \bar{p}\}| + 1 = |T| - |\{\bar{p} \in T \mid p > \bar{p}\}|. \quad (3)$$

With a related notion, Gärdenfors [14, Definition 7.3] defines the *ranking level function* to continue his discussion about modifications of the Borda function for complete preorders.

4. The following formula inspired by Hansen [16] defines the *plurality at-large* non-monotonic position operator: whenever $R \in \mathbf{W}(X)$,

$$\text{for each } x_i \in X, O_R(x_i) = \begin{cases} 0, & \text{when } x_i R x_j \text{ for each } x_j \in X, \\ 1, & \text{otherwise.} \end{cases} \quad (4)$$

Note that when $|X| > 2$, and R is such that $x_1 P x_2 P x_3 R x_k$ for each $k > 3$, we have $O_R(x_3) = O_R(x_2) = 1$ but $x_3 R x_2$ is false. Thus monotonicity is contradicted.

5. The best-worst position operator associated with $\alpha > 0$, $\beta > 0$ (inspired by the best-worst extended scoring rules studied by García-Lapresta et al. [12])

---

[3] Borda-type functions and position operators go in opposed directions. Smaller positions are associated with better options, whereas higher evaluations are given to preferred options by Borda-type functions à-la-Gärdenfors. In addition, Black [8] proves that the two functions involved in the standard definition of the Borda rule produce the same ranking.

is not monotonic either. To define it, whenever $R \in \mathbf{W}(X)$,

$$\text{for each } x_i \in X, O_R^{\alpha,\beta}(x_i) = \begin{cases} -\alpha, & \text{when } x_i R x_j \text{ for each } x_j \in X, \\ \beta, & \text{when } x_j R x_i \text{ for each } x_j \in X, \\ 0 & \text{otherwise.} \end{cases} \quad (5)$$

When $\alpha = \beta$, it produces an extended scoring rule equivalent to the dis&approval voting rule characterized by Alcantud and Laruelle [5] in the context of range or utilitarian voting. This operator has been considered for linear orders only in the aforementioned [12], and then it generates a scoring rule. In fact, García-Lapresta et al. [12] recommend dis&approval voting when voters can use complete preorders.

6. Ordinal ranks, which always break ties by assigning different positions to any pair of alternatives, violate equality [13]. One example proceeds as follows: enumerate $X = \{x_1, \ldots, x_n\}$, and in case of a tie, the order in this list acts as a tie-breaking rule.

Below, we combine the previous elements with the next standard type of aggregators:

Definition 4. *An* extended aggregation operator *is* $S: \bigcup_{m>0} \mathbb{R}^m \longrightarrow \mathbb{R}$. *It is* Paretian *if* $a_i > b_i$ *for all* $i = 1, \ldots, m$ *implies* $S(a_1, \ldots, a_m) > S(b_1, \ldots, b_m)$.

Examples of extended Paretian aggregation operators include the sum and weighted arithmetic averages, maximum, minimum, or the product if we restrict ourselves to $S: \bigcup_{m>0} \mathbb{R}_+^m \longrightarrow \mathbb{R}$.

With each position operator $\mathcal{O}$ and extended aggregation operator S, we can expand the scope of Definition 3. Indeed, for the study of profiles of complete preorders, position operators can be replaced with $S_{\mathcal{O}}: \mathcal{W}(X) \longrightarrow M(X)$ such that when $R \in \mathcal{W}(X)$, $S_{\mathcal{O}}(\mathcal{R}): M(X) \longrightarrow \mathbb{R}$ operates by $S_{\mathcal{O}}(\mathcal{R})(x) = S(O_{R_1}(x), \ldots, O_{R_m}(x))$ for each $x \in X$.

We illustrate the application of this tool with three examples.

Example 2. Consider the complete preorders R_1, R_2, R_3 on $X = \{x_1, x_2, x_3, x_4\}$ given by: $x_1 P_1 x_2 P_1 x_3 P_1 x_4$, $x_1 I_2 x_2 P_2 x_3 P_2 x_4$, and $x_2 P_3 x_1 P_3 x_3 I_3 x_4$.

To assign positions, we always use the best-worst position operator associated with $\alpha = 2$ and $\beta = 1$, denoted by $\mathcal{O}$ in this example. Its application is summarized by Table 1.

We assume that aggregation is done by sums (first case, denoted as the extended aggregation operator S_1), respectively, minimum (second case, denoted as the extended aggregation operator S_2), maximum (third case, denoted as the extended aggregation operator S_3). Definition 4 produces the following outputs, which we present in Table 2.

1. In the first case where the extended aggregation operator is S_1, option x_1 is assigned $-2-2+0$ (it is in the top-tier of the first two orders), x_2 is assigned

Table 1. A summary of the application of the best-worst position operator associated with $\alpha = 2$ and $\beta = 1$ (v., Example 1) to the complete preorders in Example 2.

	O_{R_1}	O_{R_2}	O_{R_3}
x_1	-2	-2	0
x_2	0	-2	-2
x_3	0	0	1
x_4	1	1	1

$0 - 2 - 2$ (it is in the top-tier of the second and third orders), x_3 is assigned $0+0+1$ (it is in the bottom-tier of the third order), and x_4 is assigned $1+1+1$ (it is in the bottom-tier of the three orders).

2. In the second case where the extended aggregation operator is S_2, option x_1 is assigned $\min\{-2, -2, 0\}$, x_2 is assigned $\min\{0, -2, -2\}$, x_3 is assigned $\min\{0, 0, 1\}$, and x_4 is assigned $\min\{1, 1, 1\}$.
3. In the third case where the extended aggregation operator is S_3, option x_1 is assigned $\max\{-2, -2, 0\}$, x_2 is assigned $\max\{0, -2, -2\}$, x_3 is assigned $\max\{0, 0, 1\}$, and x_4 is assigned $\max\{1, 1, 1\}$.

Table 2. Evaluations of the alternatives in Example 2 by three extended aggregation operators S_1, S_2 and S_3, respectively coupled with the best-worst position operator $\mathcal{O}$ associated with $\alpha = 2$ and $\beta = 1$.

	$S_1(\mathcal{O})$	$S_2(\mathcal{O})$	$S_3(\mathcal{O})$
x_1	-4	-2	0
x_2	-4	-2	0
x_3	1	0	1
x_4	3	1	1

We are ready to present a result that lets us establish a structured procedure that generates A-ASWFs. Below we use it to populate the list of A-ASWFs with varios proposals.

Proposition 1. *Let us fix both a monotonic position operator $\mathcal{O}$ and an extended Paretian aggregation operator S. Then $R_{S_{\mathcal{O}}(\mathcal{R})} \in \mathbf{W}_A(X)$ when $(\mathcal{R}, A) \in \mathcal{W}_A^m(X)$ and $\mathcal{R} = (R_1, \ldots, R_m) \in \mathcal{W}^A(X) \subseteq \mathcal{W}(X)$.*

Proof. For simplicity, in the rest of this proof we denote $R = R_{S_{\mathcal{O}}(\mathcal{R})}$. We need to check that xRy and $y \in A$ entail $x \in A$.

By contradiction, assume $x \notin A$. Then yP_ix for each $i = 1, \ldots, m$ because $(\mathcal{R}, A)$ is a consistent preference-approval profile. Then $O_{R_i}(x) > O_{R_i}(y)$ for each $i = 1, \ldots, m$ by monotonicity of $\mathcal{O}$. And $S(O_{R_1}(x), \ldots, O_{R_m}(x)) >$

$S(O_{R_1}(y), \ldots, O_{R_m}(y))$ follows from the fact that S is Paretian. It is now clear that xRy is false, from inspection of Eq. (2) that defines the complete preorder $R = R_{S_\mathcal{O}(\mathcal{R})}$. This contradiction ends the proof. □

Now we can use these ideas to produce the following parameterized A-ASWF in the conditions of Proposition 1, i.e., with two eligible parameters $\mathcal{O}$ and S:

$$F_\mathcal{O}^S : \mathcal{W}_A(X) \longrightarrow \mathbf{W}_A(X) \qquad (6)$$
$$(\mathcal{R}, A) \mapsto R_{S_\mathcal{O}(\mathcal{R})} \text{ where } \mathcal{R} = (R_1, \ldots, R_m)$$

For illustration, we apply this methodology to a simple case:

Example 3. (Example 2 continued). In the situation described in Example 2, we now add $A = \{x_1, x_2\}$. Then we have that (R_i, A) are preference-approval structures $(i = 1, 2, 3)$, thus $(\mathcal{R}, A) \in \mathcal{W}_A^3(X)$ with $\mathcal{R} = (R_1, R_2, R_3)$.

We proceed to apply the three A-ASWFs respectively derived from the choice of the eligible parameters of the construction that are given in Example 2. Throughout this example, $\mathcal{O}$ is the best-worst position operator associated with $\alpha = 2$ and $\beta = 1$, and S_1, S_2 and S_3 respectively denote the sum, minimum, and maximum extended aggregation operators. We draw on the conclusions presented in Table 2:

1. In the first case (namely, the extended aggregation operator S_1 coupled with $\mathcal{O}$), options x_1 and x_2 are assigned -4, x_3 is assigned 1, and x_4 is assigned 3. Therefore the operations lead to the aggregate order R such that $x_1 I x_2 P x_3 P x_4$, and indeed (R, A) is a preference-approval structure.
2. In the second case (namely, the extended aggregation operator S_2 coupled with $\mathcal{O}$), options x_1 and x_2 are assigned -2, x_3 is assigned 0, and x_4 is assigned 1. We obtain the same aggregate order than in the previous case.
3. In the third case (namely, the extended aggregation operator S_3 coupled with $\mathcal{O}$), options x_1 and x_2 are assigned 0, whereas x_3 and x_4 are assigned 1. Now we obtain the aggregate order R' such that $x_1 I' x_2 P' x_3 I' x_4$, and (R', A) is a preference-approval structure too.

Resorting to the examples listed above, we proceed to generate other particular A-ASWFs that obey to this structured construction. The next remarkable example is very popular. It contains three alternative procedures that when applied to linear orders, always coincide with the classical Borda social ordering.

Example 4. When the position operator is defined by the fractional (resp., modified, dense) rank and the aggregation operator is the sum, Eq. (6) defines the standard (resp., restricted, ranking level) Borda rule intended for complete preorders [14, Definitions 3.3, 7.1, 7.3] as explained in Eq. (2).

Example 5 below shows that we cannot dispense with monotonicity of the position operator in the statement of Proposition 1.

3.3 Incompatibilities

The following result extends Arrow's characterization of dictatorships to the setting of our work:

Theorem 1 (Santos-García and Alcantud [22]). *Pairwise dictatorial A-ASWFs are the only A-approbatory social welfare functions that satisfy unanimity and IIA.*

Now we proceed to show that other incompatibilities exist with properties that are natural in voting situations without approbatory restrictions. Indeed, social welfare functions $F' : \mathcal{W}(X) \longrightarrow \mathbf{W}(X)$ exist that satisfy the next related axioms:

1. F' satisfies Axiom T if when $x_i, x_j \in X$ are not top ranked for any of the individual orders $R_1, \ldots, R_m \in \mathbf{W}(X)$, it is the case that $x_i I x_j$ for $R = F'(\mathcal{R})$ with $\mathcal{R} = (R_1, \ldots, R_m) \in \mathcal{W}(X)$.
2. F' satisfies Axiom TB if when $x_i, x_j \in X$ are neither top nor bottom ranked for any of the individual orders $R_1, \ldots, R_m \in \mathbf{W}(X)$, it is the case that $x_i I x_j$ for $R = F'(\mathcal{R})$ with $\mathcal{R} = (R_1, \ldots, R_m) \in \mathcal{W}(X)$.

To be precise, the plurality at-large SWF satisfies Axiom T, whereas best-worst methods satisfy Axiom TB, which is weaker. Nevertheless, the transition to approbatory SWFs is less favourable, and the next example suggests why.

Example 5. Consider preference-approval structures on $X = \{x_1, x_2, x_3, x_4\}$, where $A = \{x_1, x_2, x_3\}$. In a case with $m = 2$ individuals (or 'selves') only, we take R_1, R_2, the complete preorders with $x_1 P_1 x_2 P_1 x_3 P_1 x_4$ and $x_1 I_2 x_2 P_2 x_3 P_2 x_4$. When aggregation is done by the sum, and positions are assigned by plurality at-large, recall that we cannot guarantee that the methodology derived from Proposition 1 produces A-ASWFs, because plurality at-large is not a monotonic position operator. We can nevertheless replicate the methodology.

If we routinely perform the computations, option x_1 receives two votes (it is in the top-tier of both orders), x_2 receives one vote (it is in the top-tier of the second order), and the other options receive no vote. Therefore, the aggregation produces R with $x_1 P x_2 P x_3 I x_4$ owing to Eq. (2), and we observe that (R, A) is not a preference-approval structure on X because $x_4 I x_3 \in A$ but $x_4 \notin A$.

We proceed to trace the origin of these inconveniences. In continuation of the discussion following Sect. 3.1, we say that F that satisfies Definition 2 and produces $F' : \mathcal{W}^A(X) \longrightarrow \mathbf{W}^A(X)$ satisfies Axiom T, resp., TB, when so does F'. Then one has:

Proposition 2. *There is no A-ASWF that satisfies Axiom T when $A \neq X$ and $|A| > 1$. And there is no A-ASWF that satisfies Axiom TB when $|A| > 1$ and $|X \setminus A| > 1$.*

Proof. Fix $F : \mathcal{W}_A(X) \longrightarrow \mathbf{W}_A(X)$.

In the first case, we assume without loss of generality $x_1, x_2 \in A$, $x_3 \notin A$. We use a profile with one preorder R_1 such that $x_1 P_1 x_2 P_1 x_3$ and $(R_1, A) \in \mathbf{W}_A(X)$. If F satisfies Axiom T then $x_2 I x_3$, contradicting the fact that $F(R_1, A) = (F'(R_1), A)$ is a preference-approval structure.

In the second case, we assume without loss of generality $x_1, x_2 \in A$, $x_3, x_4 \notin A$. We use a profile with one preorder R_1 such that $x_1 P_1 x_2 P_1 x_3 P_1 x_4$ and $(R_1, A) \in \mathbf{W}_A(X)$. If F satisfies Axiom TB then $x_2 I x_3$, contradicting the fact that $F(R_1, A) = (F'(R_1), A)$ is a preference-approval structure. □

4 Final Remarks

The approbatory social welfare functions defined in [22] encode the aggregation of consistent preference-approval profiles. They are suitable both for collective and multi-self choices. The starting point for this work is the limitations of existing rules for the aggregation of consistent preference-approval profiles. To overcome this omission, we have established a general procedure that guarantees a structured design for this type of aggregators. We have obtained several known generalizations of the Borda rule for complete preorders as particular cases. They all guarantee approbatory social welfare functions, thus our design can be conceived of as a generator of extensions of the Borda rule for preference-approval structures. Finally, from [22] we know that unanimity and IIA characterize the pairwise dictatorial behavior. We have shown that other properties are incompatible with the approbatory social welfare function design, albeit they are natural in popular voting systems.

Data Availability Statement. A Mathematica notebook reports the exact number of preference-approval structures, as a function of n and a, i.e., of the cardinalities of both X and the approved alternatives A. The notebook is available in a GitHub repository at https://github.com/gsantosgarcia/ApprobatorySocialWelfareFunction (accessed 2024-12-15).

Disclosure of Interests. The authors have no competing interests to declare that are relevant to the content of this article.

References

1. Albano, A., García-Lapresta, J.L., Plaia, A., Sciandra, M.: A family of distances for preference-approvals. Ann. Oper. Res. **323**(1), 1–29 (2023). https://doi.org/10.1007/s10479-022-05008-4
2. Albano, A., García-Lapresta, J.L., Plaia, A., Sciandra, M.: Clustering alternatives in preference-approvals via novel pseudometrics. Stat. Methods Appl. **33**(1), 61–87 (2024). https://doi.org/10.1007/s10260-023-00718-w
3. Alcantud, J.C.R., Cantone, D., Giarlotta, A., Watson, S.: Rationalization of indecisive choice behavior by pluralist ballots. J. Math. Econ. 102895 (2023). https://doi.org/10.1016/j.jmateco.2023.102895

4. Alcantud, J.C.R., de Andrés Calle, R., González-Arteaga, T.: Codifications of complete preorders that are compatible with Mahalanobis disconsensus measures. In: De Baets, B., Fodor, J., Montes, S. (eds.) Eurofuse 2013 Workshop. Uncertainty and Imprecision Modelling in Decision Making, pp. 19–26. Ediciones de la Universidad de Oviedo (2013)
5. Alcantud, J., Laruelle, A.: Dis&approval voting: a characterization. Soc. Choice Welfare **43**(1), 1–10 (2013). https://doi.org/10.1007/s00355-013-0766-7
6. Arrow, K.J.: Social Choice and Individual Values. Wiley (1951)
7. Ballester, M.A., Haeringer, G.: A characterization of the single-peaked domain. Soc. Choice Welfare **36**(2), 305–322 (2011). https://doi.org/10.1007/s00355-010-0476-3
8. Black, D.: Partial justification of the Borda count. Public Choice **28**(1), 1–15 (1976). https://doi.org/10.1007/BF01718454
9. Brams, S.J., Fishburn, P.C.: Approval voting. Am. Polit. Sci. Rev. **72**(3), 831–847 (1978). https://doi.org/10.2307/1955105
10. Brams, S.J., Sanver, M.R.: Voting systems that combine approval and preference. In: The Mathematics of Preference, Choice and Order: Essays in Honour of Peter C. Fishburn. Studies in Choice and Welfare, pp. 215–237. Springer (2009). https://doi.org/10.1007/978-3-540-79128-7
11. Erdamar, B., García-Lapresta, J.L., Pérez-Román, D., Remzi Sanver, M.: Measuring consensus in a preference-approval context. Inf. Fusion **17**, 14–21 (2014). https://doi.org/10.1016/j.inffus.2012.02.004
12. García-Lapresta, J.L., Marley, A., Martínez-Panero, M.: Characterizing best-worst voting systems in the scoring context. Soc. Choice Welfare **34**(3), 487–496 (2010). https://doi.org/10.1007/s00355-009-0417-1
13. García-Lapresta, J.L., Martínez-Panero, M.: Two characterizations of the dense rank. J. Math. Econ. 102963 (2024). https://doi.org/10.1016/j.jmateco.2024.102963
14. Gärdenfors, P.: Positionalist voting functions. Theor. Decis. **4**(1), 1–24 (1973). https://doi.org/10.1007/BF00133396
15. González-Arteaga, T., Alcantud, J.C.R., de Andrés Calle, R.: A new consensus ranking approach for correlated ordinal information based on Mahalanobis distance. Inf. Sci. **372**, 546–564 (2016). https://doi.org/10.1016/j.ins.2016.08.071
16. Hansen, J.A.: Comparing approval at-large to plurality at-large in multi-member districts. In: Fifth International Workshop on Computational Social Choice, Pittsburgh, PA, USA (2014)
17. Inada, K.: A note on the simple majority decision rule. Econometrica **32**(4), 525–531 (1964). https://doi.org/10.2307/1910176
18. Kalai, G., Rubinstein, A., Spiegler, R.: Rationalizing choice functions by multiple rationales. Econometrica **70**(6), 2481–2488 (2002). https://doi.org/10.1111/j.1468-0262.2002.00446.x
19. Karpov, A.: On the number of group-separable preference profiles. Group Decis. Negot. **28**(3), 501–517 (2019). https://doi.org/10.1007/s10726-019-09621-w
20. Kendall, M.G.: The treatment of ties in ranking problems. Biometrika **33**(3), 239–251 (1945). https://doi.org/10.2307/2332303
21. Santos-García, G., Alcantud, J.: Ranked soft sets. Expert. Syst. **40**(6), e13231 (2023). https://doi.org/10.1111/exsy.13231
22. Santos-García, G., Alcantud, J.: A characterization of pairwise dictatorial approbatory social welfare functions. Econ. Lett. **248**, 112217 (2025). https://doi.org/10.1016/j.econlet.2025.112217

Evaluating the eNego System: A Dual Perspective on Subjective Acceptation and Objective Scoring System Accuracy

Tomasz Wachowicz[1] and Ewa Roszkowska[2(✉)]

[1] University of Economics in Katowice, 1 Maja 50, 40-287 Katowice, Poland
tomasz.wachowicz@uekat.pl
[2] Bialystok University of Technology, Wiejska 45A, 15-351 Bialystok, Poland
e.roszkowska@pb.edu.pl

Abstract. This study aims to analyze the evaluation of use and usefulness of the eNego electronic negotiation system from both subjective and objective perspectives. The subjective evaluation is based on responses to questionnaire items inspired by selected elements of the Technology Acceptance Model, specifically perceived ease of use and perceived usefulness. The objective evaluation stems from the analysis of the accuracy of scoring systems created by negotiators in eNego. To achieve this goal, we utilized experimental data from the series of eNego experiments. The analysis employs descriptive statistical methods and k-means clustering to identify distinct user profiles that capture similar subjective evaluations of eNego, as well as to explore the relationship between these profiles and the scoring system's accuracy. The results provide valuable insights into user preferences and the overall effectiveness of the system. The analysis of user interactions with the eNego system revealed varying evaluations across four user groups, determined by k-cluster analysis. Initial inaccuracies in the negotiation scoring system were improved through an iterative pre-negotiation mechanism, ultimately leading to comparable accuracy levels across all groups.

Keywords: eNego · subjective and objective evaluation · ease of use · usefulness · negotiation scoring system · accuracy · user profile

1 Introduction

Negotiation is a complex process in which two or more parties with differing interests and preferences seek to reach a mutually acceptable agreement [1, 2]. This complexity often involves analyzing multiple options, evaluating trade-offs, and making decisions under time pressure. To assist with these tasks, e-Negotiation Systems (eNSs) have been developed. These computer-based tools support negotiators at various stages of the negotiation process, from preparation to execution, and help analyze agreements.

Kersten and Lai [3] provide a historical overview of software designed to support negotiations and automate tasks. One of the first online negotiation systems, Inspire [4], was developed for studying intercultural negotiations in educational contexts. Inspire offered tools for gathering preferences and generating scoring systems used for offers evaluation, suggesting alternatives through a search-based offer generator, analyzing incoming offers, and evaluating the efficiency of the final agreement. It also tracked negotiation progress using history graphs and provided an easily navigable transcript for users. The solutions pioneered in Inspire served as inspiration for modern systems, such as Negoisst [5], SmartSettle [6], Web Hipre [7], TOBANS [8], and eNego [9].

ENSs often incorporate multiple criteria decision-aiding (MCDA) techniques to build negotiation scoring systems and provide decision support to negotiators. A key challenge is ensuring that when providing such support, eNSs should account for the cognitive limitations and abilities of users. When eNSs are properly aligned with the negotiators' needs, they are more likely to be accepted and used effectively. User-system interaction is influenced by the negotiator's experience, skills, and intuition, and any complexity or lack of understanding of the system can hinder its effectiveness [10–12]. Unfortunately, most prenegotiation protocols implemented in negotiation software to facilitate decision-making rely on MCDA techniques based on the preference aggregation approach, which assumes that negotiators can easily express their preferences on an atomic level (e.g., weights, options) numerically. These preferences are then aggregated to evaluate negotiation offers. Experimental studies show that despite its simplicity, direct rating mechanisms used there have limitations due to negotiators' cognitive constraints. These constraints make it difficult to handle abstract numbers and accurately map preference information into the scoring system [9, 12–15]. To address such issues, the eNego[1] system was designed in which an alternative approach based on the preference disaggregation-aggregation paradigm was adopted [9, 16]. This holistic approach is considered less cognitively demanding, as it asks decision-makers to compare alternatives that are more realistic to the decision problem at hand, making it more natural than evaluating abstract options without context [17].

The goal of this study is to evaluate the acceptance of the eNego system from both subjective and objective perspectives. The subjective evaluation uses the responses to questionnaire items inspired by inventories used in the Technology Acceptance Model (TAM) [18–21] to evaluate perceived ease of use (PEU) and perceived usefulness (PU). They allow us to assess how intuitive and effective the system is for users. The objective evaluation investigates the improvements and accuracy of the negotiation scoring system generated through eNego's pre-negotiation module. Data from a questionnaire and an online negotiation experiment were analyzed using statistical methods and k-means clustering to explore user preferences and the system's effectiveness. The results provide insights into how users with different profiles of satisfaction (measured by a combination of PEU and PU ratings) interact with the system. Furthermore, the findings demonstrate that the system's accuracy improved over time through an iterative tuning mechanism.

The paper is organized into three more sections. Section 2 provides an overview of the eNego system, while Sect. 3 outlines the methodology for its evaluation. Section 4 presents the empirical study on user acceptance and scoring system accuracy in the eNego system. The paper concludes with a summary of key findings.

---

[1] https://webs.ue.katowice.pl/enego/.

2 ENego Negotiation Support System: An Overview

In line with Kersten-Lai taxonomy [3], eNego can be categorized as an e-Negotiation System for research and training. Similar to the Inspire system, it offers three key support functions: (1) structuring the process into distinct phases and activities, (2) eliciting preferences and constructing rating functions, and (3) visualizing the progress of the negotiation. Additionally, eNego incorporates modules for data collection and user surveys. These features are actively utilized in research to analyze negotiation profiles, evaluate user interactions in various negotiation scenarios, and assess system effectiveness (for a detailed description, see [9]).

It facilitates bilateral, synchronous, or asynchronous multi-issue negotiations and supports both pre-negotiation, bargaining, and postnegotiation activities. Controlled by an administrator, eNego can be customized to meet experimental or teaching requirements by configuring a negotiation protocol, which outlines the steps and activities for users. The system uses a database of predefined negotiation templates assigned to experiments by the administrator. ENego supports individual negotiation template evaluation through various coded modules. During negotiations, users can build and exchange offers, send messages, and view the negotiation history.

ENego employs a hybrid holistic procedure for pre-negotiation preference elicitation, combining disaggregation and aggregation methods. The UTA algorithm [22] is applied for disaggregation, while MARS (Measuring Alternatives near Reference Solutions) [23] is used to determine reference alternatives. Additional tools manage nonmonotonic preferences during pre-negotiation. The holistic procedure implemented in eNego consists of four structured steps designed to elicit and refine negotiators' preferences effectively [9, 16]:

1. **Calibrating Preference Monotonicity**: Defining the best and worst options for each issue to confirm the monotonicity of preferences.
2. **Ranking Reference Offers**: Organizing MARS-based reference alternatives in descending order of preference, utilizing a simple drag-and-drop interface for clarity and ease.
3. **Tuning Template Ratings:** Adjusting scores determined from a modified UTA model. Ratings are refined iteratively, and inconsistencies are resolved to ensure coherent alignment of preferences.
4. **Aggregating and Verifying Scores:** Analyzing global scores of selected offers and verifying that the rankings reflect the true quality of preferences. Further refinements are made as needed to ensure accuracy and user satisfaction.

The above procedure can be performed iteratively. After each step of preference disaggregation (before step 3), eNego informs negotiators of the optimization process results. If the disaggregation mechanism fails to identify at least one set of marginal utility functions reflecting the preferences from step 2, the negotiator is notified. In such cases, they are advised to repeat steps 1 and 2 or pay special attention during step 3. Current experiments conducted in eNego utilize a bilateral multi-issue negotiation scenario inspired by the original Cypress-Itex negotiations from the Inspire system [4]. The negotiation involves two parties: a bicycle producer and a parts supplier, negotiating a contract for rear-wheel gear delivery. Four issues are negotiated: price, delivery time, payment time, and returns, with predefined options forming the negotiation template. Students act as agents representing principals. During pre-negotiation, they review the case description and confidential preferential information, which detailed the principals' goals and priorities. This information is visualized using pie charts, aiding in the preference elicitation process. eNego negotiators first complete the preference elicitation process, aiming to accurately reflect their principal's priorities, and then enter the bargaining phase to reach a mutually satisfying agreement.

3 Methodology for Evaluating the eNego System

The evaluation of satisfaction from the eNego system is conducted from two key perspectives: subjective and objective. The subjective evaluation focuses on users' individual experiences with the system, including aspects such as ease of use and usefulness. In contrast, the objective evaluation relies on measurable parameters related to the negotiation scoring system's accuracy and improvements.

The subjective evaluation of eNego acceptance is based on data collected through an online questionnaire comprising seven questions. All items were rated on a 7-point Likert scale (1 = strongly agree, 7 = strongly disagree) and were inspired by the notions initially defined in the Technology Acceptance Model (TAM), where some questions addressed Perceived Ease of Use (PEU), and others focused on Perceived Usefulness (PU) [20]. In eNego, PEU-related items evaluate how straightforward, manageable, and free from unnecessary complexity the system's preference elicitation and ranking processes are, as well as how time-consuming they might be. They are:

- **V1:** The entire preference elicitation process in eNego was cumbersome and time-consuming.
- **V2:** It was difficult for me to build a ranking using the predefined alternatives.
- **V3:** The set of predefined alternatives was too numerous.

These statements capture users' perceptions of the system's intuitiveness, user-friendliness, and potential barriers to efficient usage.

PU-related items capture the extent to which users believe the eNego system effectively supports their negotiation tasks and aligns with their preferences and needs. They are:

- **V4:** If I had the option, I would use a different set of alternatives for comparison.
- **V5:** The interface that required dragging and dropping boxes with offer examples was unintuitive and inefficient.
- **V6:** The scoring system determined by the UTA procedure did not accurately reflect my principal's preferences.
- **V7:** I would prefer to assign the issue and option ratings myself just from the very beginning, without any preceding holistic declarations of rankings of exemplary offers.

The above items reflect users' perceptions of the system's effectiveness, relevance, and ability to adapt to their negotiation needs.

The objective evaluation is based on the data recorded by eNego system independently from students' option provided in the postnegotiation questionnaires. This data describes the user's actions taken during the prenegotiation phase when interacting with eNego interface. It assesses measurable attributes of the negotiation scoring system built by the user in the eNego system, such as the number of changes made to the initial scoring system generated by the UTA tool and the improvement in the accuracy of this scoring system. The evaluation is based on three variables:

- **Number of Changes** (NC): The number of modifications made to the initial UTA-generated negotiation scoring system, including manual adjustments to issue scores or changes made in ranking alternatives through the user interface (number of repetitions of steps 2 and 3).
- **Accuracy First** (AF): The accuracy of the scoring system initially generated by users through UTA-based holistic algorithm.
- **Accuracy Last** (AL): The accuracy of the final scoring system after all modifications made by users in prenegotiation.

To measure the accuracy of the scoring system, we compared the systems built by eNego users to the reference scoring systems of their principals, which corresponded precisely to the description of preferences provided in the negotiation instructions. Each agent's scoring system was compared to their principal's using accuracy metrics. Cardinal accuracy measures the discrepancies in the strength of preferences between the agent's and the principal's scoring systems, using Manhattan distance. This involves calculating the absolute differences in the ratings of all options and summing them up. According to the characteristics of the AF and AL indexes, lower values indicate higher accuracy (for details see [9, 16]).

The procedure for analyzing negotiations in eNego, as described in this study, follows a three-step approach:

Step 1: Basic Statistical Analysis of Variables

The analysis begins by examining the frequency distributions of rating items V1–V7. A chi-square test will be conducted to explore potential relationships between pairs of these items. For the variables NC, AF, and AL, basic descriptive statistics—including the minimum, maximum, mean, and standard deviation—will be calculated to provide a comprehensive overview of the improvements and accuracy of the negotiation scoring system.

Step 2: Identifying User Profiles Using the k-Means Method

Responses to items V1–V7 will be analyzed using the k-means clustering method to identify user profiles based on their subjective evaluation of the eNego system. These profiles will be described using descriptive statistical measures, and the Mann-Whitney test will be applied to assess the significance of the differences between them.

K-means clustering was selected due to its well-established position in multivariate statistical analysis, with numerous applications and favorable properties regarding stability and convergence [24]—particularly in linear spaces, which constitute the typical interpretive framework for TAM-based item responses. Furthermore, some of the method's well-known limitations, such as the risk of converging to suboptimal solutions, can be effectively mitigated by employing a multistart procedure with sufficiently diversified initial centroids.

Step 3: Analysis of the Relationship Between User Profiles and Negotiation Scoring System Accuracy

The Mann-Whitney test was employed to examine the relationship between identified profiles of users and the accuracy of the negotiation scoring system. The analysis will explore whether user profiles influence scoring system accuracy and how the system's adaptive improvement mechanisms addressed initial inaccuracies. Additionally, the Wilcoxon test will be used to analyze the extent of improvement in the profiles.

4 Empirical Study

4.1 Experimental Setup

The online negotiation experiment was conducted within the eNego system, allowing participants to engage in a preference elicitation and ranking process. The UTA-based tool, utilizing the MARS approach [9, 16], generated an initial scoring system based on users' preferences. Participants were encouraged to modify this initial scoring to enhance its accuracy and relevance. The online questionnaire, consisting of items V1–V7, was designed to assess users' subjective perceptions of the eNego system, while the objective evaluation focused on the variables NC, AF, and AL. This dual approach enabled the assessment of both subjective and objective aspects of the eNego system by evaluating user experiences, preferences, and the performance of the negotiation scoring system.

The experiments within the eNego system were conducted between 2018 and 2024. These studies involved students from six Polish universities, resulting in 994 valid records after data cleaning. The students engaged with eNego as part of their coursework, obtaining an adequate training in domain of negotiation and decision-making, and to enhance the consequentiality of their participation, their performance within the system was incorporated into their final course grades. Among the participants, 61.6% were male and 38.4% were female. Participants' ages ranged from 18 to 50, with a mean age of 23 years (SD = 4.25).

4.2 Basic Statistical Analysis of Variables

Figure 1 presents a heatmap illustrating the distribution of responses to items V1–V7, providing an overview of how participants responded to questionnaire items.

The distribution of responses to the items V1-V7 reveals key trends in user perceptions. Over 30% of participants gave high scores (6 or 7) for each item, reflecting generally positive views on various aspects of the eNego system. Conversely, less than 15% assigned low scores (1 or 2), indicating more negative assessments of the system. In particular, almost one-third felt the scoring system accurately reflected their principal's preferences (V6), while more than half found the drag-and-drop interface relatively intuitive and efficient (V5). In contrast, nearly 15% preferred to assign the issue and option ratings themselves (V7).

Next, the Chi-square test of independence was applied to examine the association between responses for each pair of items. The results revealed statistical significance ($p < 0.001$) for all pairs, indicating a strong relationship between responses to them. This suggests that users' ratings for one item were significantly related to their ratings of others, implying that various aspects of the eNego system were perceived as interconnected by the participants.

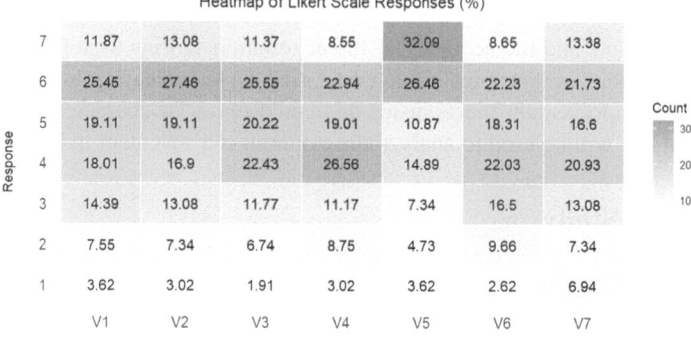

Fig. 1. The heatmap with values representing the percentages of responses on the Likert scale for items V1–V7.

The mean AF value is 117.48 (SD = 57.17), with a range from 34.77 to 463.60. Higher AF values indicate greater initial inaccuracies in the system, necessitating improvement. Perfectly accurate systems have an AF of 0, while the average AF for randomly generated scoring systems is approximately 240 (see [16]). Among the eNego users, 3.1% (30 negotiators) produced scoring systems less accurate than the average random system, which may reflect either carelessness in declaring preferences (low engagement) or a misunderstanding of the concept of generating scoring systems from ordered examples of offers. The NC variable shows that users made between 1 and 9 modifications to their negotiation scoring systems, with a mean of 3.10 (SD = 1.90), reflecting moderate interaction with the system. These modifications, intended by the designers of the pre-negotiation protocol, aimed to refine preferences and produce scoring systems that better represented users' priorities. Indeed, the mean inaccuracy of scoring systems after iterative improvements (AL) is 79.27 (SD = 50.19), with a range from 10.7 to 410.20. This demonstrates that modifications generally improved system accuracy, as evidenced by the lower AL values compared to AF. The difference is statistically significant, confirmed by the Wilcoxon test ($p < 0.001$). However, the system still did not achieve optimal accuracy for many users, suggesting areas for further refinement to better align the system with users' needs.

4.3 Identifying User Profiles Using the k-Means Method

In the next step of the analysis, we examined whether the responses to items V1-V7 could form distinct clusters of user profiles. Using the k-means clustering method, we identified four distinct profiles based on participants' evaluations of the eNego system. Figure 2 presents the mean values for items V1-V7 across these profiles. To assess the differences between groups, the Mann-Whitney test was applied. Statistically significant differences ($p < 0.003$) were observed for all pairs of items, except for V4 for G2-vs-G3 ($p = 0.68$) and V5 for G2-vs-G4 ($p = 0.05$). Let us note that in the second case, the p-value was marginally insignificant, indicating a borderline result.

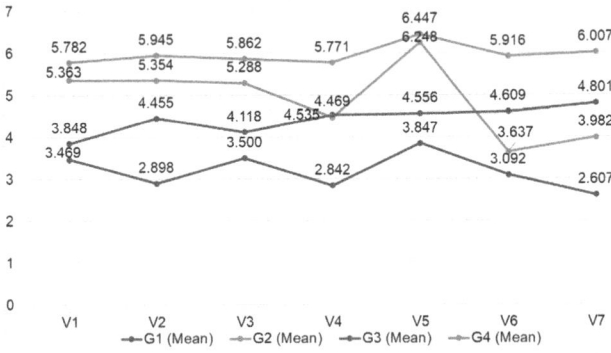

Fig. 2. The mean values for items V1–V7 across users' profiles

To provide a clearer understanding of profiles and more effectively characterize the differences between them, Fig. 3 presents the box plots for items V1–V7 across the groups. These box plots offer a detailed visual representation of the distribution of responses within each group, allowing for deeper insight into the variations and tendencies in the users' evaluations of the eNego system.

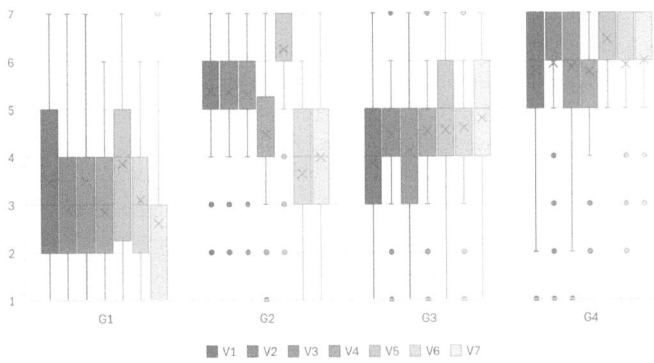

Fig. 3. The box plots for items V1–V7 across the groups

Group 1 (N = 196) consistently reports the lowest average ratings across all items, with scores ranging from 2.607 to 3.847. This group demonstrates moderate dissatisfaction with the eNego system, Interquartile Spans (IQSs) for difficulty in ranking alternatives (V2), number of predefined alternatives (V3), preference for alternative comparison sets (V4), and alignment of scoring with user's principal preferences fall between 2 and 4, underscoring significant challenges in these aspects. Broader IQRs of 2 to 5 for time-consuming and difficulty preference elicitation (V1) and drag-and-drop interface (V5) indicate that these features were particularly problematic for many respondents. Meanwhile, the narrower IQR of 1 to 3 for self-assigned ratings (V7) reveals a unified dissatisfaction with UTA-based scoring. Overall, users in this group appear less convinced of the system's value, potentially finding it difficult to engage with effectively.

In stark contrast, Group 4 (N = 275) demonstrates consistently high ratings, with mean scores ranging from 5.771 to 6.447, reflecting strong satisfaction across all aspects of the eNego system. IQRs for V1 and V3 span 5 to 7, while V4 has a slightly narrower IQR of 5 to 6, indicating a consistently positive evaluation. The remaining items exhibit IQRs, ranging from 6 to 7, reinforcing the group's overall strong satisfaction. Respondents in this group perceive the system as effective, engaging, and easy to use.

Groups 2 and 3 occupy intermediate positions, with average ratings falling between those of Groups 1 and 4, reflecting more balanced or moderate views of the system. Key differences between those groups emerge upon closer comparison.

In Group 2 (N = 226), mean scores range from 3.637 to 6.248, reflecting a generally favorable assessment of the system's usability and interface. Group 2 reports higher scores for ease-of-use-related items, with IQRs spanning 5 to 6, such as time-consuming and difficulty in preference process (V1), difficulty in ranking alternatives (V2), and the number of predefined alternatives (V3), suggesting that these respondents find the system easier to navigate. Additionally, Group 2 assigns very high ratings to interface intuitiveness and effectiveness (V5), with an IQR of 6 to 7, nearly matching the levels reported by Group 4, emphasizing the importance of the system's intuitive design However, evaluations of usefulness (V6 and V7) are more moderate, with IQRs from 3 to 5, suggesting that while users find the system easy to use, they are less confident in its effectiveness and flexibility.

For Group 3 (N = 297), the mean scores range from 3.848 to 4.401, indicating slightly higher satisfaction than Group 1 but still moderate compared to Group 4. Item related to time-consuming and difficulty preference elicitation (V1) and the number of predefined alternatives (V3) have IQRs spanning from 3 to 5, while usability of drag-and-drop interface (V5) and self-assigned ratings (V7) range from 4 to 6. For the remaining items (V2, V4, V6), IQRs range from 4 to 5. Group 3 scores higher than Group 2 on the accuracy of the UTA-based scoring system (V6) and preference for self-assigned ratings (V7). These results highlight that Group 3 focuses on the scoring system's accuracy, with a more balanced view of ease of use, reflecting moderate satisfaction across most items. Both groups, G1 and G2, gave similar evaluations regarding their preference for alternative comparison sets (V4), with interquartile ranges (IQRs) between 4 and 5. No significant differences were found for this item. This suggests that Group 3 places a higher importance on scoring accuracy, preferring more control over the process. Their views on ease of use are more balanced, with ratings indicating moderate satisfaction across most items.

Based on these observations, the groups can be categorized as follows:

- Group 1: The "**Less Satisfied Users**" – characterized by low (IQRs 1–3 or 2–4) and disparities (IQRs 2–5) ratings across all items, reflecting a general dissatisfaction with the system.
- Group 2: The "**Ease of Use and Interface-Focused Users**" – these users prioritize system usability (IQRs 3–5), with a strong preference for ease of use (IQRs 5–6) and user-friendly design (IQRs 6–7).

- Group 3: The "**Usefulness-Focused Functional Users**"– these respondents emphasize the importance of scoring accuracy with user-friendly design (IQRs 4–6 or 4–5) with a moderate view of the system's ease of use (IQRs 3–5 or 4–5).
- Group 4: The "**Enthusiastic Users**"– exhibiting consistently high ratings, indicating strong satisfaction with all aspects of the system (IQRs 5–7, 6–7, or 5–6).

4.4 Analysis Relationships Between Users Profiles and Negotiation Scoring System Accuracy

Finally, we explored the relationships between the subjective evaluation of the eNego system's perceived ease of use and perceived usefulness and the objective assessment of its utility expressed through the variables NC, AF, and AL. The analysis examines whether significant differences in these characteristics (NC, AF, AL) exist across the four user groups (G1–G4). Table 1 presents the mean values of variables NC, AF, and AL across the groups.

Table 1. Mean values for variables NC, AF, and AL across the groups

Score	Least Satisfied (G1)	Ease of Use and Interface-Focused (G2)	Usefulness-Focused Functional (G3)	Enthusiastic (G4)
NC	3.26	3.45	2.92	2.90
AF	133.78	113.34	119.35	107.28
AL	87.59	75.62	79.60	75.99

Firstly, it is important to recognize that inaccuracies in negotiation scoring may stem from both difficulties in using the eNego system and from user limitations or cognitive biases [12–14]. In this analysis, we focus exclusively on the eNego system itself.

The "Least Satisfied Users" group (G1) demonstrated the highest level of inconsistency in the initial scoring system generated by UTA, with an average Accuracy First (AF) score of 133.78. The Mann-Whitney test revealed a statistically significant difference ($p < 0.01$) for AF between G1 and the other groups (G2, G3, G4). The poor performance of G1 can be attributed to difficulties interacting with the system and the interface, which could have led to greater inaccuracies in the scoring.

In contrast, the "Enthusiastic Users" (G4) group exhibited, on average, the smallest inaccuracy AF equal to 107.28, reflecting their ease of interaction with the system. Respondents in G4, who rated the eNego system highly, encountered minimal difficulties using the system for negotiation scoring. The Mann-Whitney test also showed a statistically significant difference ($p < 0.01$) between G4 and G3, but no significant difference when comparing G2 and G4 ($p = 0.104$). Meanwhile, the "Ease of Use and Interface-Focused Users" (G2) and "Usefulness-Focused Functional Users" (G3) groups exhibited moderate on average inaccuracy AF with 113 for G2 and 119 for G3, but these differences were statistically insignificant (Mann-Whitney test, $p = 0.072$), suggesting their challenges with the system were less pronounced than those of G1, although still noteworthy.

As a result of the modifications made, significant improvements in the system's consistency were achieved within each group. On average, the largest number of changes were made by respondents in G2 (mean NC = 3.45), followed by G1 (mean NC = 3.26), although the observed differences in NC between these groups were statistically insignificant (Mann-Whitney test, $p = 0.453$). On the other hand, the fewest modifications were observed in G4 (mean NC = 2.90) and G3 (mean NC = 2.92). Statistically significant differences in NC were found (Mann-Whitney test, $p < 0.001$) between G1 and G4, G2 and G4, and G2 and G3. This suggests that dissatisfaction, previously attributed to the poor quality of the initial scoring systems, may also have been influenced by a higher intensity of interaction with the preference analysis module in these groups compared to others.

It is important to note that regardless of the group, the ability to manually refine the negotiation scoring system resulted in measurable benefits across all groups. This suggests that this feature of the eNego system was highly effective. In each group, the mechanism for improving scoring was successfully used to enhance the accuracy of the system (Wilcoxon text; $p < 0.001$), resulting in the best final accuracy in G2 and G4 (mean AL equal to 75.62 and 75.99, respectively, with an insignificant difference in the Mann-Whitney test, $p = 0.574$). The accuracy was somewhat lower in G3 (mean AL = 79.60) and the lowest in G1 (mean AF = 87.59). However, after adjustments were made, there were no significant differences in AL between any two groups (Mann-Whitney test, $p > 0.192$).

5 Conclusion

Based on the analysis of user interactions with the eNego system, several key findings emerge. These findings reflect the varied experiences of different user groups and highlight both subjective and objective perspectives in the evaluation of the eNego system.

The study identified distinct groups of respondents who evaluated the ease of use and usefulness of the eNego system differently. At one end of the spectrum, group G1 consistently gave low ratings across all aspects of the system, while at the other end, respondents in group G4 rated the system highly on all dimensions. Between these extremes, two moderate groups emerged, differing in their evaluations of ease of use and usefulness. These groups appeared to favor either the system's user-friendliness or its usefulness more strongly than the other.

A relationship was evident between the initial inaccuracy and respondents' evaluations of the system: those who rated the eNego system more positively demonstrated statistically significantly better accuracy in the initial negotiation scoring system compared to those who rated the system's ease of use and usefulness poorly. Similarly, the two moderate groups, which were moderately satisfied with either the ease of use or the usefulness of the system, displayed intermediate levels of initial scoring inaccuracies.

The pre-negotiation protocol introduced in eNego, allowing iterative improvement of the scoring system, proved highly effective. Despite initial inaccuracies, the system's adaptability facilitated significant improvements across all user groups. Users from the group that rated the system lowest (G1) made the largest number of adjustments, likely because they required more effort to achieve a level of accuracy comparable to other groups. Conversely, users who rated the system most enthusiastically (G4) made the fewest changes, suggesting that they either recognized the system's functionality more intuitively or required fewer adjustments to improve its performance.

As a result of applying this improvement mechanism, the final negotiation scoring system showed notable improvements. Interestingly, small differences in the average accuracy of the final scoring systems were observed between groups, which were statistically insignificant across all groups of eNego users, indicating that all groups ultimately achieved similar levels of accuracy of the negotiation scoring system after the adjustments.

In light of the above conclusions, it is important to acknowledge the limitations of this study stemming from the adopted research assumptions and methodological choices. The first issue concerns the use of students in experimental analyses. The usefulness of student samples is a widely debated issue in empirical economics [25], information systems [26] and negotiation research [27] with no definitive conclusions reached. Herbst and Schwarz [27] found that students who received negotiation training outperformed their untrained peers and were not significantly outperformed by professional negotiators. This suggests that trained students can be effectively used as experimental subjects in negotiation research. In our case, the sample comprised students who had undergone training in decision-making and negotiation as part of their academic coursework involving the use of eNego. Therefore, we believe their domain knowledge was comparable to that of managers who deal with negotiation and contracting.

The notions of ease of use and usefulness applied in this study were initially introduced in the TAM model by Davis in 1989 [21]. While TAM remains a widely used model for studying technology adoption, other models employ similar evaluation constructs, including the refined Innovation Diffusion Theory (IDT) [28], as well as both the original and modified versions of the Unified Theory of Acceptance and Use of Technology (UTAUT [29] and extended UTAUT2 [30]). Although the concepts used to evaluate eNego in this study are shared—albeit differently labeled and classified—across these models, it is important that future research aiming to comprehensively assess eNego adopts a more holistic framework. Such a framework should capture all potential interdependencies, including the influence of the evaluation of the preference analysis module(s) on the overall assessment of the system, as well as a range of external variables—for example, those used in this study to describe users' objective experiences and outcomes from using the system.

Our future research will focus on developing such more advanced causal models linking subjective and objective evaluations of the eNego system's usefulness with other elements of the extended TAM or UTAUT2 framework, such as external and context-specified factors. Additionally, we aim to include qualitative data from text analysis in the evaluation of the system's functionality. eNego users were also asked to provide open-ended feedback on the system and its modules, offering valuable insights that could further enhance its design and usability.

Comparative studies evaluating similar electronic negotiation environments (eNEs) would also be of interest, as they could reveal the extent to which the overall system evaluation depends on its decision support components. These components may be more or less explicitly visible within the system, depending on the interface design solutions adopted in different platforms. In particular, the use of AI-based modules that support the process of constructing such evaluations could significantly influence users' perceptions of the system.

Acknowledgment. This research was supported by the 'Regional Initiative of Excellence' program financed by the Polish Ministry of Science and Higher Education and the grant WZ/WI-IIT/2/2025 from Bialystok University of Technology.

References

1. Thompson, L.: The Mind and Heart of the Negotiator. Pearson/Prentice Hall Upper Saddle River, NJ, New Jersey (2005)
2. Raiffa, H., Richardson, J., Metcalfe, D.: Negotiation Analysis: The Science and Art of Collaborative Decision Making. Harvard University Press (2002)
3. Kersten, G.E., Lai, H.: Negotiation support and e-negotiation systems: an overview. Group Decis. Negot. **16**, 553–586 (2007). https://doi.org/10.1007/s10726-007-9095-5
4. Kersten, G., Noronha, S.J.: WWW-based negotiation support: design, implementation, and use. Decis. Support. Syst. **25**, 135–154 (1999). https://doi.org/10.1016/S0167-9236(99)00012-3
5. Schoop, M., Jertila, A., List, T.: Negoisst: a negotiation support system for electronic business-to-business negotiations in e-commerce. Data Knowl. Eng. **47**, 371–401 (2003)
6. Thiessen, E.M., Soberg, A.: Smartsettle described with the montreal taxonomy. Group Decis. Negot. **12**, 165 (2003)
7. Mustajoki, J., Hämäläinen, R.P.: Web-Hipre: global decision support by value tree and AHP analysis. INFOR: Inf. Syst. Oper. Res. **38**, 208–220 (2000). https://doi.org/10.1080/03155986.2000.11732409
8. Wachowicz, T., Błaszczyk, P.: TOPSIS based approach to scoring negotiating offers in negotiation support systems. Group Decis. Negot. **22**, 1021–1050 (2013). https://doi.org/10.1007/s10726-012-9299-1
9. Wachowicz, T., Roszkowska, E.: Holistic preferences and prenegotiation preparation. In: Kilgour, D.M., Eden, C. (eds.) Handbook of Group Decision and Negotiation, pp. 255–289. Springer, Cham (2021)
10. Roszkowska, E., Wachowicz, T.: Towards cognitive decision support: a model of behavioural assessment of multi-criteria methods. Control. Cybern. **50**, 145–168 (2021). https://doi.org/10.2478/candc-2021-0009
11. Kersten, G., Cray, D.: Perspectives on representation and analysis of negotiation: towards cognitive support systems. Group Decis. Negot. **5**, 433–467 (1996). https://doi.org/10.1007/BF02404644
12. Kersten, G., Roszkowska, E., Wachowicz, T.: How well agents represent their principals' preferences: the effect of information processing, value orientation, and goals. In: Szapiro, T., Kacprzyk, J. (eds.) Collective Decisions: Theory, Algorithms And Decision Support Systems, pp. 119–151. Springer International Publishing, Cham (2022). https://doi.org/10.1007/978-3-030-84997-9_6
13. Kersten, G., Roszkowska, E., Wachowicz, T.: The heuristics and biases in using the negotiation support systems. In: Schoop, M., Kilgour, D.M. (eds.) Group Decision and Negotiation. A Socio-Technical Perspective, pp. 215–228. Springer International Publishing, Cham (2017). https://doi.org/10.1007/978-3-319-63546-0_16
14. Kersten, G., Roszkowska, E., Wachowicz, T.: An impact of negotiation profiles on the accuracy of negotiation offer scoring system? Exper. Study. Multiple Criteria Decis. Making **11**, 77–103 (2016)

15. Kersten, G., Roszkowska, E., Wachowicz, T.: Representative decision-making and the propensity to use round and sharp numbers in preference specification. In: Chen, Y., Kersten, G.E., Vetschera, R., Xu, H. (eds.) Group Decision and Negotiation in an Uncertain World. GDN 2018, pp. 43–55. Springer, Cham (2018). https://doi.org/10.1007/978-3-319-92874-6_4
16. Roszkowska, E., Wachowicz, T., Kersten, G.: Can the holistic preference elicitation be used to determine an accurate negotiation offer scoring system? A comparison of direct rating and UTASTAR techniques. In: Schoop, M., Kilgour, D.M. (eds.) Group Decision and Negotiation. A Socio-Technical Perspective, pp. 202–214. Springer International Publishing, Cham (2017). https://doi.org/10.1007/978-3-319-63546-0_15
17. Wachowicz, T., Roszkowska, E.: Can holistic declaration of preferences improve a negotiation offer scoring system? Eur. J. Oper. Res. **299**, 1018–1032 (2022). https://doi.org/10.1016/j.ejor.2021.10.008
18. Corrente, S., Greco, S., Kadziński, M., Słowiński, R.: Robust ordinal regression in preference learning and ranking. Mach. Learn. **93**, 381–422 (2013)
19. Legris, P., Ingham, J., Collerette, P.: Why do people use information technology? A critical review of the technology acceptance model. Inf. Manage. **40**, 191–204 (2003). https://doi.org/10.1016/S0378-7206(01)00143-4
20. Davis, F.D.: User acceptance of information technology: system characteristics, user perceptions and behavioral impacts. Int. J. Man Mach. Stud. **38**, 475–487 (1993)
21. Davis, F.D.: A technology acceptance model for empirically testing new end-user information systems. Theory and Results/Massachusetts Institute of Technology (1986)
22. Davis, F.D.: Perceived usefulness, perceived ease of use, and user acceptance of information technology. MIS Q., 319–340 (1989)
23. Siskos, Y., Grigoroudis, E., Matsatsinis, N.F.: UTA methods. In: Greco, S., Ehrgott, M., Figueira, J.R. (eds.) Multiple Criteria Decision Analysis: State of the Art Surveys, pp. 297–334. Springer (2016). https://doi.org/10.1007/978-1-4939-3094-4_9
24. Górecka, D., Roszkowska, E., Wachowicz, T.: The MARS approach in the verbal and holistic evaluation of the negotiation template. Group Decis. Negot. **25**, 1097–1136 (2016). https://doi.org/10.1007/s10726-016-9475-9
25. Ahmed, M., Seraj, R., Islam, S.M.S.: The k-means algorithm: a comprehensive survey and performance evaluation. Electronics **9** (2020). https://doi.org/10.3390/electronics9081295
26. Levitt, S.D., List, J.A.: What do laboratory experiments measuring social preferences reveal about the real world? J. Econ. Perspect. **21**, 153–174 (2007)
27. Compeau, D., Marcolin, B., Kelley, H., Higgins, C.: Research commentary—generalizability of information systems research using student subjects—a reflection on our practices and recommendations for future research. Inf. Syst. Res. **23**, 1093–1109 (2012). https://doi.org/10.1287/isre.1120.0423
28. Herbst, U., Schwarz, S.: How valid is negotiation research based on student sample groups? New insights into a long-standing controversy. Negot. J. **27**, 147–170 (2011). https://doi.org/10.1111/j.1571-9979.2011.00300.x
29. Rogers, E.M.: Diffusion of Innovations: Modifications of a Model for Telecommunications. In: Stoetzer, M.W., Mahler, A. (eds.) Die Diffusion von Innovationen in der Telekommunikation. pp. 25–38. Springer, Berlin, Heidelberg (1995). https://doi.org/10.1007/978-3-642-79868-9_2
30. Venkatesh, V., Morris, M.G., Davis, G.B., Davis, F.D.: User acceptance of information technology: toward a unified view. MIS Q. **27**, 425–478 (2003). https://doi.org/10.2307/30036540
31. Schretzlmaier, P., Hecker, A., Ammenwerth, E.: Extension of the unified theory of acceptance and use of technology 2 model for predicting mHealth acceptance using diabetes as an example: a cross-sectional validation study. BMJ Health Care Inform. **29**, e100640 (2022). https://doi.org/10.1136/bmjhci-2022-100640

Conflict Modeling in Complex Decision Environments

Redefining Sequential Stability for Multi-decision-Maker Conflict Within Graph Model

Ziming Zhu[1,2](✉) [iD], D. Marc Kilgour[3,4] [iD], and Keith W. Hipel[3] [iD]

[1] International College, Guangzhou College of Commerce, Guangzhou 511363, China
zzm08@tsinghua.org.cn
[2] Research Centre for Contemporary Management, Tsinghua University, Beijing 100084, China
[3] Department of Systems Design Engineering, University of Waterloo, Waterloo, ON N2L 3G1, Canada
[4] Department of Mathematics, Wilfrid Laurier University, Waterloo, ON N2L 3C5, Canada

Abstract. When a two-decision-maker conflict is generalized to a multi-decision-maker conflict, definitions of sequential stability must be refined because of ambiguities in meaning of sanctions. In an n-decision-maker conflict ($n > 2$), when a focal decision maker/coalition has a unilateral improvement, its opponents will construct a sequence of movements to sanction the focal decision maker/coalition. A sanction is believable if it is preferred by all the sanctioners. After analyzing the problems of the existing definitions of non-cooperative and cooperative sequential stabilities within the paradigm of graph model, convincing definitions for sanctions of a focal decision maker/coalition's initial unilateral improvements are proposed.

Keyword: Graph Model · Conflict Analysis · Sequential Stability Definitions · Convincing Sanction

1 Introduction

When they generalized a two-player game to an n-player game, von Neumann and Morgenstern [1] thought that the challenge was analyze possible coalitions. A coalition is a group of players cooperating for their mutual benefit. The players making up a coalition are distinct and each have their own preferences.

The Graph Model for Conflict Resolution (GMCR) provides a flexible and operational methodology for analyzing a conflict from a non-cooperative standpoint [2, 3]. GMCR contains a rich range of stability definitions that provides the analyst with a clear picture of how the conflict can evolve, and how the decision makers (DMs)' decision styles and behaviors are reflected in its evolution. These stability definitions consist of Nash [4, 5], general metarational [6], symmetric metarational [6], sequential stability [7, 8], limited-move stabilities [3, 9], non-myopic stability [3, 9–11], symmetric sequential stability (SSEQ) [12], higher-order sequential stability [13], and time sensitive stabilities [14]. In this paper, we focus only on sequential stability, partially because it is regarded as one of most useful, and easiest to apply, definitions [15].

A graph model is called non-cooperative if commitments are not enforceable, even if pre-play communication between the DMs is possible. It is called cooperative if commitments – agreements, promises, threats – are fully binding and enforceable. Correspondingly, two types of sequential stability exist.

The emphasis in the non-cooperative theory is on the individual, on what strategy should be chosen in one's own interest. For n-DM non-cooperative conflicts, sequential stability was originally put forward by Fraser & Hipel [16, 17]. Fang et al. [15] reformulated it for the graph model and determined its relations to other stability definitions.

Within the framework of GMCR, n-DM non-cooperative stability concepts have been expanded to deal with a variety of cooperative behavior situations, leading naturally to the concept of coalitions. The different approaches to cooperative analyses constitute models of joint actions by members of a coalition, and of how a coalition can change the outcomes of a strategic conflict. There are two main approaches: one is to aggregate the preferences of coalition members according to some metric and treat the coalition as a single new DM; the other assumes an independent DM will not change his or her preferences but tries, as the conflict evolves, to find common and achievable improvements with other coalition members.

The first class of coalition analysis within the framework of GMCR has been implemented using both state-based and option-based metrics. Kuhn et al. [18] took a state-based perspective, developing a metric for the similarity of preferences among the members of a proposed coalition as an indicator of the likelihood of the formation of this coalition. Hipel and Meister [19] provided an option-based metric related to the preference tree, in which a DM's preferences are expressed in terms of options. In both the state and option-based metric approaches, it is assumed that more similar DMs are more likely to form a coalition. Next, the preferences of the individual DMs contained in a coalition are aggregated using a simple, but sensible, procedure using either the state-based or option-based approach. For a coalition containing DMs with very different preferences, the combined preferences may contain more equally preferred states.

The second class of coalition analysis within the GMCR paradigm considers that an individual DM's preferences remain unchanged within any coalition it may join. Specifically, Kilgour et al. [20] suggested why and when independent DMs may act together as a coalition and developed a procedure for determining which non-cooperative equilibria could be threatened by a coalition of DMs. Subsequently, Inohara and Hipel [21] expanded these ideas by defining coalitional stabilities within the GMCR framework. Inohara and Hipel [22] established theorems regarding interrelationships among coalitional and non-cooperative stability concepts. Xu et al. [23] expanded coalition stability to include uncertain preference using a matrix representation within the framework of GMCR. Other sequential definitions of coalition equilibria have been proposed by Xu et al. (Chapter 8) [24]. All of the above research depends upon having a classical coalition improvement, an idea that was extended to a Pareto coalition improvement by Zhu et al. [25] to incorporate the assumption that a coalition would not form unless it would benefit at least one member and harm none. The ability to carry out coalition analysis in multi-party conflicts made the inverse graph model a hot research area in recent years [26–31]. The development of coalitional analyses within the graph model has included some novel definitions, such as Pareto stabilities [32], the full coalition set [33], and new interrelationships among stabilities [34].

In n-DM real-world conflicts, there are usually many sequentially stable equilibria, depending on how sanctioning improvements are redefined or refined. If there are fewer sanctioning states, then the number of equilibria will decrease. To refine the idea of sanctions, we suggest four rules as follows:

- In a sanction, everyone in the sanction should prefer this sanction to the original UI.
- Every move from the original UI to the sanction should be a UI for the mover.
- The sanctioning coalition should contain only DMs whose assistance is required to reach the sanction. (Infinite loops are excluded under this condition.)
- Have it more demanding to have a sanction, so there are not too many equilibria.

The goal of this paper is to refine non-cooperative and cooperative sequential stability definitions within the paradigm of GMCR. The main contribution of this paper is to refine definitions of equilibria in an n-DM-conflict by defining novel coalitional improvements within the paradigm of the GMCR.

The structure of the remainder of this paper is as follows. Section 2 provides the reasons of refining a series of existing n-DM sequential stabilities. Section 3 gives a range of refined sequential definitions under both non-cooperative and cooperative circumstances. Finally, conclusions are drawn and further research suggested in Sect. 4.

2 Why Redefine n-DM Sequential Stability?

2.1 Basic Notations of GMCR

Basically, a graph model contains [35]:

1) A finite set N of DMs, $N = \{1, 2, \cdots, n\}$;
2) A finite set S of feasible states;
3) For each $i \in N$, a directed graph, $G_i = (S, A_i)$. [The state set S is the set of vertices of G_i. The set of arcs is $A_i \subseteq S \times S$; A_i is the set of all unilateral moves controlled by DM i. It is assumed that the directed graph G_i has no loops or multiple edges.]
4) For each $i \in N$, a strongly complete set of relations $\{\succ_i, \sim_i\}$ on S. [For any $s_1, s_2 \in S$, $s_1 \succ_i s_2$ means that DM i prefers state s_1 to s_2, and $s_1 \sim_i s_2$ means that DM i is indifferent between state s_1 and s_2. DM i's preference relation, $\succ_i$, must be irreflexive, asymmetric and transitive, and DM i's indifference relation, $\sim_i$, must be reflexive, asymmetric and transitive. Strong completeness means that, for any $s_1, s_2 \in S$, exactly one of $s_1 \succ_i s_2$, $s_1 \prec_i s_2$, and $s_1 \sim_i s_2$ must be true.]

For any state $s \in S$, let $\Phi_i^{\preceq}(s)$ be the set of all states that are less preferred or equally preferred to state s by DM i. Hence, $\Phi_i^{\preceq}(s) = \{s_1 \in S | s \succ_i s_1 \text{ or } s \sim_i s_1\}$.

2.2 State Transitions and Sanctions

Within the graph model, a state transition is a move from one state to another, controlled by a specific DM.

Definition 1 (Unilateral Movement, UM). For each $i \in N$ and $s \in S$, a unilateral move of DM i is any $s_1 \in S$ such that $(s, s_1) \in A_i$.

The concepts of reachable list and unilateral improvement (UI) list of a DM often used in stability definitions and the evolution of a conflict, are given as follows.

Definition 2 (Reachable List). For each $i \in N$ and each $s \in S$, DM i's reachable list from state s, $R_i(s) = \{s_1 \in S | (s, s_1) \in A_i\} \subset S$.

$R_i(s)$ is the set of all states that DM i can reach (in one step) from state s, i.e., $R_i(s)$ consists of all those states s_1 such that there is an arc from s to s_1 in $G_i = (S, A_i)$.

Definition 3 (Unilateral Improvement List). For each $i \in N$ and each $s \in S$, DM i's unilateral improvement list from state s, $R_i^+(s) \subseteq R_i(s)$, consists of those states $s_1 \in R_i(s)$ that are preferred to s. Formally, $R_i^+(s) = \{s_1 \in R_i(s) | s_1 \succ_i s\}$.

If $s_1 \in R_i^+(s)$, then i can move directly from s to s_1 and i prefers s_1 to s. In this case, s_1 is called a unilateral improvement (UI) for i from s. It follows that $R_i^+(s) \subseteq R_i(s)$.

When there are more than two DMs in a dispute, the other DMs, given by $N - \{i\}$, where $|N - \{i\}| \geq 2$, may execute a sequence of moves to sanction a UI by DM i. Hence, sanctioning DMs in $N - \{i\}$ act like members of a coalition, even though each move is made non-cooperatively (We consider that a coalition is a set of DMs acting together to achieve results which are desirable for all DMs in the set [12]). In general, let $H \subseteq N$ represent a coalition of DMs participating in a conflict. The coalition H is non-empty if $|H| > 0$, trivial if $|H| = 1$ and non-trivial if $|H| > 1$. Two kinds of allowable moves by coalition members are defined below.

Definition 4 (Unilateral Move of Coalition). Let $s \in S$, $H \subseteq N$, and $H \neq \emptyset$. A unilateral movement by a coalition H is a member of $R_H(s) \subseteq S$, which is defined inductively as follows. Initially, define $\Omega : S \to 2^N$ by $\Omega(s_1) = \emptyset$ for all $s_1 \in S$. Now carry out the following steps as long as possible:

(1) If $i \in H$ and $s_1 \in R_i(s)$, then $s_1 \in R_H(s)$ and $\Omega(s_1) = \Omega(s_1) \cup \{i\}$;

(2) If $i \in H$, $s_1 \in R_i(s)$, $\Omega(s_1) \neq \{i\}$, and $s_2 \in R_H(s_1)$, then $s_2 \in R_H(s)$ and $\Omega(s_2) = \Omega(s_2) \cup \{i\}$.

Definition 4A (Equivalent Definition of Unilateral Move of Coalition). Let $s \in S$, $H \subseteq N$, and $H \neq \emptyset$. The unilateral movement list of H from s is $R_H(s) = \{t \in S :$ there exists $s_0, s_1, s_2, \cdots, s_k \in S$ such that $s_0 = s$, $s_k = t$, and $s_j \in R_{i_j}(s_{j-1})$ for $j = 1, 2, \cdots, k$, where $i_j \in H$ and, for $j > 1$, $i_j \neq i_{j-1}\}$.

Similar to $R_H(s)$ above, the definition of individual unilateral improvement list of coalition H from state s, denoted by $R_H^+(s) \subseteq R_H(s)$, is given as follows:

Definition 5 (Individual Unilateral Improvement of Coalition). Let $s \in S, H \subseteq N$, and $H \neq \emptyset$. An individual unilateral improvement by H is a member of $R_H^+(s) \subseteq S$, which is defined inductively. Initially, define $\Omega^+ : S \to 2^N$ by $\Omega^+(s_1) = \emptyset$ for all $s_1 \in S$. Now carry out the following steps as long as possible:

(1) If $i \in H$ and $s_1 \in R_i^+(s)$, then $s_1 \in R_H^+(s)$ and $\Omega^+(s_1) = \Omega^+(s_1) \cup \{i\}$;

(2) If $i \in H$, $s_1 \in R_i^+(s)$, $\Omega(s_1) \neq \{i\}$, and $s_2 \in R_H^+(s_1)$, then $s_2 \in R_H^+(s)$ and $\Omega^+(s_2) = \Omega^+(s_2) \cup \{i\}$.

Definition 5A (Equivalent of Individual Unilateral Improvement of Coalition). Let $s \in S$, $H \subseteq N$, and $H \neq \emptyset$. The individual unilateral improvement list of H from s is $R_H^+(s) = \{t \in S : \text{there exists } s_0, s_1, s_2, \cdots, s_k \in S \text{ such that } s_0 = s, s_k = t, \text{ and } s_j \in R_{i_j}^+(s_{j-1}) \text{ for } j = 1, 2, \cdots, k, \text{ where } i_j \in H \text{ and, for } j > 1, i_j \neq i_{j-1}\}$.

2.3 Non-cooperative Sequential Stability

Definition 6 (Sequential Stability). For $i \in N$, a state $s \in S$ is sequentially stable for DM i, denoted $s \in S_i^{SEQ}$, iff for every $s_1 \in R_i^+(s)$, there exists $s_2 \in R_{N-i}^+(s_1) \cap \Phi_i^{\preceq}(s)$.

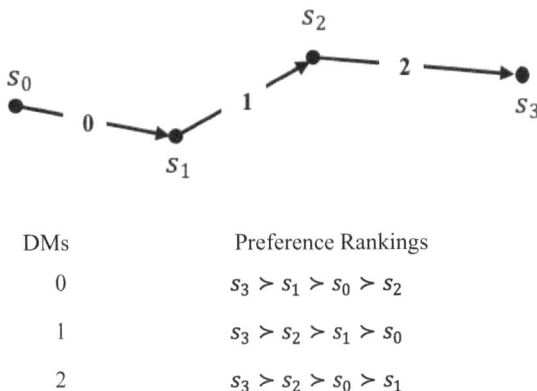

DMs	Preference Rankings
0	$s_3 \succ s_1 \succ s_0 \succ s_2$
1	$s_3 \succ s_2 \succ s_1 \succ s_0$
2	$s_3 \succ s_2 \succ s_0 \succ s_1$

Fig. 1. Example 1 of an Integrated Graph Model with 3 DMs and 4 States

Comment: In Example 1 (Fig. 1), is s_0 sequentially stable for DM 0? Obviously, the sanctioning coalition might include some or all of the DMs other than DM 0. However, the sanction of an individual UI by $N - H$ may not be reasonable. To illustrate, note that there is an initial UI from s_0 by DM 0 and $s_1 \in R_0^+(s_0)$, but the other DMs could form a coalition $N - \{0\} = \{1, 2\}$ to sanction this UI. Note that $R_{N-0}^+(s_1) = \{s_2, s_3\}$ and $s_0 \succ_0 s_2$, so state s_0 is SEQ stable for DM 0 according to Definition 6. But a contrary view is that s_2 is a **temporary** or **intermediate** state in the sequence of unilateral improvements by $N - \{0\}$ starting from state s_1 in the sense that at least one DM (other than the one who moved last) has at least one more UI. Note that there is no binding agreement among the members of $N - \{0\}$ about who will act or when to stop. Consequently, at state s_2, DM 2 might seize the initiative and move to state s_3, rather than stay at state s_2, because $s_3 \succ_2 s_2$. So, this UI sequence stops at state s_3. And $s_3 \succ_0 s_0$ and DM 0 knows the graph and the other DM's preferences. Consequently, DM 0 might not regard state s_2 as a sanction. Moreover, s_3 satisfies another condition: it makes every member of the coalition better off. Therefore, state s_0 is not SEQ stable for DM 0, establishing a contradiction between the SEQ definition and the intuition for a graph model with n DMs ($n > 2$).

2.4 Coalitional Sequential Stability Definitions

Kilgour et al. [20] put forward the classical coalition improvement concept and applied it to interpret the coalition form in the well-known Elmira conflict example.

Definition 7 (Classical Coalition Improvement) [20]. For a non-empty coalition $H \subseteq N$ and $s \in S$, the coalition improvement list of coalition H from state s is $CR_H^+(s) = \{s_1 \in R_H(s) | \forall i \in H, s_1 \succ_i s\}$.

Thus, if $s_1 \in CR_H^+(s)$, then s_1 is reachable by H from s, and every member of H prefers s_1 to s. Inohara & Hipel [21] generalized the concept of Kilgour et al. [20] to define a series of stabilities using Definitions 8–10. First, if $M \subseteq N$ and $M \neq \emptyset$, then $\mathbb{P}(M)$ is the subclass of all nonempty coalitions within M, i.e., $\mathbb{P}(M) = \{H \subseteq M : M \neq \emptyset\}$.

Definition 8 (Class Reachable List of Subclass) [21]. Let $\emptyset \neq M \subseteq N$. For a subclass $\mathbb{P}(M)$ and $s \in S$, the class reachable list of $\mathbb{P}(M)$ from s is defined inductively as the set $R_{\mathbb{P}(M)}(s)$ that two conditions: (i) If $H \in \mathbb{P}(M)$ and $t \in R_H(s)$, then $t \in R_{\mathbb{P}(M)}(s)$, and (ii) if $H \in \mathbb{P}(M)$, $t \in R_{\mathbb{P}(M)}(s)$, and $u \in R_H(t)$, then $u \in R_{\mathbb{P}(M)}(s)$.

Definition 9 (Class Improvement List of Subclass) [21]. Let $\emptyset \neq M \subseteq N$. For a subclass $\mathbb{P}(M)$ and $s \in S$, the class improvement list of subclass $\mathbb{P}(M)$ from state s is defined inductively as the set $CR_{\mathbb{P}(M)}^+(s)$ that satisfies two conditions: (i) If $H \in \mathbb{P}(M)$ and $t \in CR_H^+(s)$, then $t \in CR_{\mathbb{P}(M)}^+(s)$, and (ii) If $H \in \mathbb{P}(M)$, $t \in CR_H^+(s)$ and $u \in CR_H^+(s)$, then $u \in CR_{\mathbb{P}(M)}^+(s)$.

For any state $s \in S$, let $\Phi_H^{\preceq}(s)$ be the set of all states that are not more preferred than state s for at least one DM in coalition H, that is $\Phi_H^{\preceq}(s) = \{x \in S | \exists i \in H, s \succcurlyeq_i x\}$. Obviously, $\Phi_H^{\preceq}(s) = \bigcup_{i \in H} \Phi_i^{\preceq}(s)$.

Definition 10 (Coalitional Sequential Stability for a Coalition) [21]. For $H \in \mathbb{P}(N)$, state $s \in S$ is coalitional sequentially stable for coalition H, denoted by $s \in S_H^{CSEQ}$, if and only if for all $s_1 \in CR_H^+(s)$, $CR_{\mathbb{P}(N-H)}^+(s_1) \cap \Phi_H^{\preceq}(s) \neq \emptyset$.

Comment: For the coalitional SEQ stability definitions given by [21, 22], each coalition behaves like a DM. Moreover, the sanctioning coalition(s) might include some or all of the DMs. So, there is same problem as discussed of the Definition 6. In Example 1, state s_0 is coalition sequentially stable because of the possible move by DM 1 to s_2, as $s_2 \succ_1 s_1$ and $s_2 \prec_0 s_0$.

In their coalitional SEQ stability definitions, Xu et al.[24] saw coalition H's opponents as either the entire coalition $N - H$, or the individual DMs in $N - H$, but not any intermediate coalition. Thus, if $|N - H| > 2$, some subcoalitions of opponents are missing. They are accounted for in Definitions 11 and 12.

Definition 11 (Coalitional Sequential Stability with Whole Coalition Sanctions). Let $H \subseteq N$ be a nonempty coalition. State $s \in S$ is *coalitional sequentially stable* (CSEQ$_1$) for H, denoted by $s \in S_H^{CSEQ_1}$, iff for every $s_1 \in CR_H^+(s)$, there exists $s_2 \in CR_{N-H}^+(s_1) \cap \Phi_H^{\preceq}(s) \neq \emptyset$.

Comment: For the CSEQ$_1$ stabilities, coalition H's opponents $N - H$ are treated as a whole coalition. A sanctioning UI must be preferred by all DMs in $N - H$. Note that some SEQ stabilities might be lost because sanctions by subcoalitions of $N - H$ are not allowed. In Example 2 (Fig. 2), state s_0 is not CSEQ$_1$ stable for $H = \{0\}$ according to Definition 11, because facing $s_1 \in R_0^+(s_0)$, $CR_{\{1,2\}}^+(s_1) = \varnothing$ for a whole sanctioning coalition $N - H = \{1,2\}$. However, facing a UI by DM 0, $s_1 \in R_0^+(s_0)$, DM 1 could make a UI from s_1 to s_2, and $s_2 \prec s_0$, so s_0 is obviously sequentially stable.

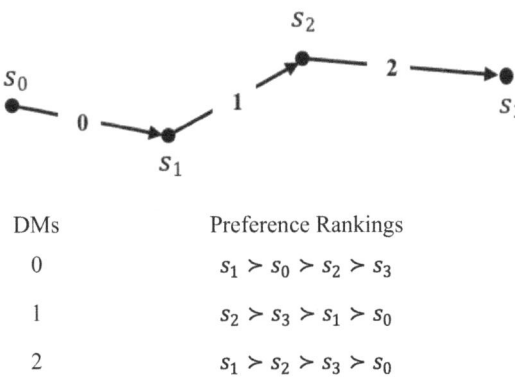

DMs	Preference Rankings
0	$s_1 \succ s_0 \succ s_2 \succ s_3$
1	$s_2 \succ s_3 \succ s_1 \succ s_0$
2	$s_1 \succ s_2 \succ s_3 \succ s_0$

Fig. 2. Integrated Graph of Example 2, a Graph Model with 3 DMs and 4 States

Definition 12 (Coalitional Sequential Stability Sanctioned by Individual DMs). Let $H \subseteq N$ be a nonempty coalition. State $s \in S$ is coalitional sequential stable (CSEQ$_2$) for H, denoted by $s \in S_H^{CSEQ_2}$, iff for every $s_1 \in CR_H^+(s)$, there exists $s_2 \in R_{N-H}^+(s_1) \cap \phi_H^{\preceq}(s) \neq \varnothing$.

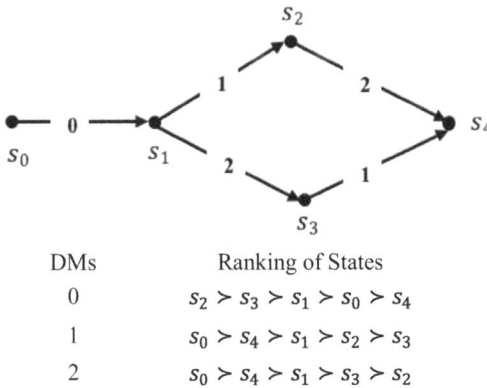

DMs	Ranking of States
0	$s_2 \succ s_3 \succ s_1 \succ s_0 \succ s_4$
1	$s_0 \succ s_4 \succ s_1 \succ s_2 \succ s_3$
2	$s_0 \succ s_4 \succ s_1 \succ s_3 \succ s_2$

Fig. 3. Integrated Graph of Example 3, a Graph Model with 3 DMs and 5 States

Comment: Under $CSEQ_2$, only individual members of $(N - H)$ may sanction $H's$ coalition improvements. So, the sanctioning moves are unilateral improvements by $N - H$ (just as in the non-cooperative graph model) the initial move is a coalitional UI. Obviously, these definitions are inconsistent.

In Example 3 (Fig. 3), state 0 is not sequentially stable according to Definition 12, because $R_{N-H}^+(s_1) = \varnothing$ for $N - H = \{1, 2\}$. But, facing a UI by DM 0, $s_1 \in R_0^+(s_0)$, there exists a coalitional improvement $s_4 \in CR_{\{1,2\}}^+(s_1)$ by coalition $\{1, 2\}$, and $s_4 \prec s_0$. So s_0 is coalitionally sequentially stable for DM 0. Moreover, s_0 is coalitional sequential stable for DM 1 and 2. So, in the sense of Definition 12, Example 3 s_0 exhibits coalitional SEQ stability.

3 Redefine Sequential Stability Definitions for n-DM Graph Model

Previously, the resolution concepts for an n-DM conflict model are divided into two groups: individual/non-cooperative stability definitions and coalitional stability definitions. But in an n-DM conflict model, the individual UI by a DM might be sanctioned by one or more coalitions. The n-DM Graph model could be classified as either individual/non-cooperative or cooperative. In the individual/non-cooperative graph model, every DM acts individually and has no binding agreements with any other DM. In the former, only individual UIs are considered. In the latter, DMs can cooperate and achieve initial coalitional UIs (of course, including initial individual UIs) that are preferred to outcomes available through individual actions.

3.1 Individual/non-Cooperative Sequential Stability Definitions for n-DM Graph Model

In the Definitions 13–20, a refinement of coalition improvement is put forward, and then integrated into a definition of coalitional sequential stability.

Definition 13 (Net Unilateral Improvement of Coalition). Let $s \in S, H \subseteq N$, and $H \neq \varnothing$. The net unilateral improvement list of coalition H from state s, is defined as the set $\{t \in R_H(s) | t \succ_i s, for all \, i \in H\} = R_H^{+,A}(s)$.

Comment: A net unilateral improvement is same as the coalitional improvement in Definition 7. From the initial state to the terminal state on a UI sequence, there might exist one or more intermediate states that make the mover worse off temporally, so a binding commitment is needed among the coalition members. If a net unilateral improvement is regarded as a sanction, then the SEQ definition is given as follows.

Definition 14 (Sequential Net Stability). For $i \in N$, a state $s \in S$ is sequentially net stable for DM i, denoted $s \in S_i^{SEQ,A}$, iff for every $s_1 \in R_i^+(s)$, there exists $s_2 \in R_{N-i}^{+,A}(s_1) \cap \Phi_i^{\preceq}(s)$.

Comment: The initial UI is made individually, but the sanctioning UI is made by a coalition with a binding commitment. A net UI of a coalition makes sure all DMs are better off, i.e., that the terminal state preferred by each DM in coalition H. Actually, the definition of a net UI for a coalition is same as the classical coalitional improvement of a coalition, defined by Inohara and Hipel [21]. However, in the process of unilateral movement to the terminal state, some transitory states might be not preferred by some DMs in coalition H. So binding commitment among all the coalition members is required. Clearly, the initial and sanctioning UIs can be inconsistent in this regard.

Definition 15 (Consistent Unilateral Improvement of Coalition). Let $s \in S, H \subseteq N$, and $H \neq \emptyset$. The consistent unilateral improvement list of coalition H from state s, is defined as the set $\{t \in R_H(s) |\ there exists s_0, s_1, s_2, \cdots, s_k\ such that s_0 = s, s_k = t, and s_j \in R^+_{i_j}(s_{j-1}) for all j = 1, \cdots, k\} = R^{+,B}_H(s)$.

Definition 16 (Sequential Consistent Stability). For $i \in N$, a state $s \in S$ is sequentially consistently stable for DM i, denoted $s \in S_i^{SEQ,B}$, iff for every $s_1 \in R_i^+(s)$, there exists $s_2 \in R^{+,B}_{N-i}(s_1) \cap \Phi_i^{\preceq}(s)$.

Definition 17 (Consistent Net Unilateral Improvement of Coalition). Let $s \in S$, $H \subseteq N$, and $H \neq \emptyset$. The consistent net unilateral improvement list of coalition H from state s, is defined as, $R^{+,C}_H(s) = R^{+,A}_H(s) \cap R^{+,B}_H(s)$.

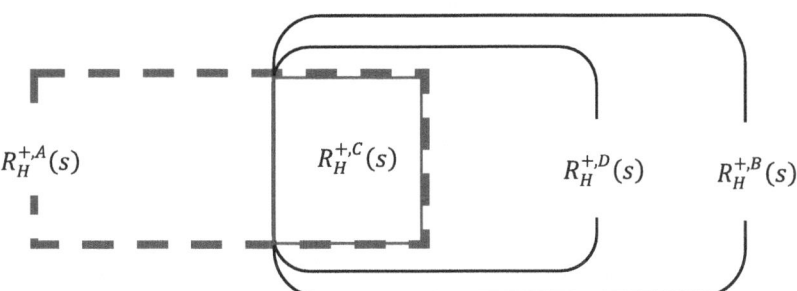

Fig. 4. Relationships among UIs A-D of Coalition Improvements

Definition 18 (Sequential Consistent Net Stability). For $i \in N$, a state $s \in S$ is sequentially consistent net stable for DM i, denoted $s \in S_i^{SEQ,C}$, iff for every $s_1 \in R_i^+(s)$, there exists $s_2 \in R^{+,C}_{N-i}(s_1) \cap \Phi_i^{\preceq}(s)$.

Comment: A consistent net UI of a coalition is both a consistent UI and a net UI of the coalition. So, it is preferred by all the DMs in a coalition and does not need a binding agreement among the all the DMs in the coalition. But some other possible stabilities might be omitted.

Definition 19 (Solid Unilateral Improvement of Coalition). Let $s \in S, H \subseteq N$, and $H \neq \emptyset$. A solid unilateral improvement of coalition H from state s is a member of $R^{+,D}_H(s) = \{t \in R_H(s) | there exists s = s_0, s_1, s_2, \cdots, s_k = t, s_j \in R^+_{i_j}(s_{j-1}) for all j = 1, \cdots, k, and t \succ_i s for all i = i_j, for j = 1, 2, \cdots, k\}$.

Here, s_k is called the end state of the UI sequence. Note that some DMs in H may not contribute to the individual UI sequence, but are included because they prefer s_k to s_0, or cannot move at all. Obviously, $R_H^{+,C}(s)$ is contained by $R_H^{+,D}(s)$ according to Definitions 17 and 19, and $R_H^{+,D}(s)$ is contained by $R_H^{+,B}(s)$ according to Definitions 15 and 19 (See Fig. 4).

Definition 20 (Solid Sequential Stability). For $i \in N$, a state $s \in S$ is solidly sequentially stable for DM i, denoted $s \in S_i^{SEQ,D}$, iff for every $s_1 \in R_i^+(s)$, there exists $s_2 \in R_{N-i}^{+,D}(s_1) \cap \Phi_i^{\preceq}(s)$.

Comment: Every DM involved in the UI sequence prefers s_k to s_1, making the sanction solid.

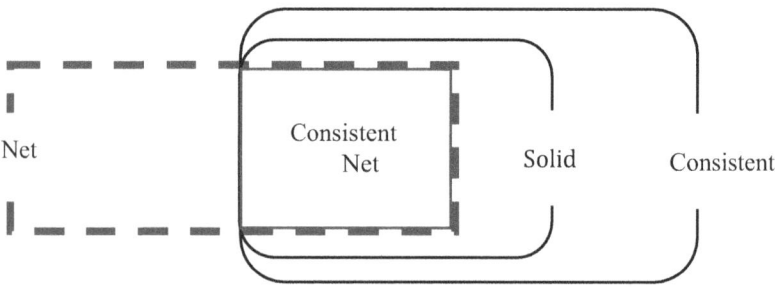

Fig. 5. Relationships of different sequential stabilities

If the initial improvements are fixed, then more sanctions mean more stabilities. Accordingly, the relationships of the four kinds of stability are the same as for the UIs (See Fig. 5).

3.2 Cooperative Sequential Stability Definitions for n-DM Graph Model

For the cooperative graph model, the initial coalitional UI is important. The idea of Pareto coalitional UI, put forward by Zhu et al. [25], is more general than the UIs which we have just defined.

Definition 21 (Pareto Coalition Improvement) [25]. For a non-empty coalition $H \subseteq N$ and $s, s_1 \in S$, s_1 is a Pareto coalition improvement for coalition H from state s, denoted $s_1 \in PR_H^+(s)$, if and only if:

(1) $s_1 \in R_H(s)$, and.

(2) At least one DM $i \in H$ strictly prefers state s_1 to s (that is, $s_1 \succ_i s$, for some $i \in H$) and all DMs in H, weakly prefer state s_1 to state s (that is, $s_1 \succcurlyeq_j s$ for $j \in H$).

Definition 22 (Solid Class Pareto Improvement of Subclass). For a subclass of $\mathbb{C}$ of $\mathbb{P}(N)$ and $s \in S$, $s_1 \in S$ is a solid class Pareto improvement of subclass $\mathbb{C}$ from state s, In this case, $s_1 \in PR_{\mathbb{C}}^{+,D}(s) \subseteq S$. Let $s \in S, H \subseteq N$, and $H \neq \emptyset$.

Definition 23 (Solid Pareto Sequential Stability). For a non-empty coalition $H \subseteq N$ and $s \in S$, a state $s \in S$ is Pareto sequentially stable for H, denoted $s \in S_i^{PSEQ.D}$, iff for every $s_1 \in PR_H^+(s)$, there exists $s_2 \in PR_{\mathbb{P}(N-H)}^{+,D}(s_1) \cap \Phi_H^{\preceq}(s)$.

Compared to the classical coalitional sequential stabilities by Inohara and Hipel[37], when the initial improvements increase, the stabilities decrease; however, as the sanctioning improvements increase, the stabilities increase. In the end, either an increase or a decrease is possible.

4 Conclusions and Further Research

In an n-DM conflict model within the Graph Model paradigm, both the existing non-cooperative and cooperative sequential stabilities might rely on sanctions that are not solid. We have analyzed the situation and proposed some refined sequential stability definitions.

We maintain that a DM/coalition's behavior should be **consistent** for both initial and sanctioning improvements. For an n-DM ($n > 2$) graph model, the initial improvement behavior should be consistent with the sanctioning improvement. A DM/coalition might attend both initial improvements and sanctioning improvements. Secondly, the sanctioning improvement sequence should be **solid**. That is to say, a sanctioning improvement state should be preferred by all DMs who attend this improvement sequence. Thirdly, no sanctioning improvement state should be **missed**. Every possible sanctioning improvement should be considered for, otherwise, some sequential stabilities might be neglected.

No existing sequential definition can satisfy these three standards. In this paper, we have proposed refined non-cooperative and cooperative sequential stabilities in which DMs/coalitions' improvement behaviors do meet all three standards.

We recognize that, in the process of making a coalitional improvement -- whether initial or sanctioning -- there might form a loop or loops. They will be studied in a future paper. Additionally, matrix formulation [24] of all of the foregoing improved sequential stability definitions would be useful.

Acknowledgments. The authors thank the three anonymous referees for their insightful comments and suggestions. The second author gratefully acknowledges the financial support from Natural Sciences and Engineering Research Council (NSERC) Discovery grant (No. RGPIN-2019–5903). The third author gratefully acknowledges the financial support from Natural Sciences and Engineering Research Council (NSERC) Discovery grant (No. RGPIN-2018–4345).

Disclosure of Interests. The authors have no competing interests to declare that are relevant to the content of this article.

References

1. Von Neumann, J., Morgenstern, O.: Theory of Games and Economic Behavior. Princeton University Press, Princeton, NJ, USA (1953)

2. Fang, L., Hipel, K.W., Kilgour, D.M.: Interactive Decision Making: The Graph Model for Conflict Resolution. John Wiley & Sons, New York, NY, USA (1993)
3. Kilgour, D.M., Hipel, K.W., Fang, L.: The graph model for conflicts. Automatica **23**, 41–55 (1987)
4. Nash, J.: Equilibrium points in n-person games. Proc. Nat. Acad. Sci. USA **36**, 48–49 (1950)
5. Nash, J.: Non-cooperative games. Ann. Math. **54**, 286–295 (1951)
6. Howard, N.: Paradoxes of Rationality: Theory of Metagames and Political Behavior. MIT press, Cambridge, MA, USA (1971)
7. Fraser, N.M., Hipel, K.W.: Solving complex conflicts. IEEE Trans. Syst. Man Cybern. **9**, 805–816 (1979)
8. Fraser, N.M., Hipel, K.W.: Conflict Analysis: Models and Resolutions. North-Holland, New York, NY, USA (1984)
9. Kilgour, D.M.: Anticipation and stability in two-person noncooperative games. Dyn. Models Int. Conflict, 26–51 (1985)
10. Brams, S.J., Wittman, D.: Nonmyopic Equilibria in 2 × 2 Games. Confl. Manag. Peace Sci. **6**, 39–62 (1981). https://doi.org/10.1177/073889428100600103
11. Kilgour, D.M.: Equilibria for far-sighted players. Theory Decis. **16**, 135–157 (1984). https://doi.org/10.1007/BF00125875
12. Rêgo, L.C., Vieira, G.I.A.: Symmetric sequential stability in the graph model for conflict resolution with multiple decision makers. Group Decis. Negot. **26**, 775–792 (2017). https://doi.org/10.1007/s10726-016-9520-8
13. Rêgo, L.C., de Oliveira, F.E.G.: Higher-order sequential stabilities in the graph model for conflict resolution for bilateral conflicts. Group Decis. Negot. **29**, 601–626 (2020). https://doi.org/10.1007/s10726-020-09668-0
14. He, S.: A time sensitive graph model for conflict resolution with application to international air carbon negotiation. Eur. J. Oper. Res. **302**, 652–670 (2022). https://doi.org/10.1016/j.ejor.2022.01.019
15. Fang, L., Hipel, K.W., Kilgour, D.M.: Conflict models in graph form: solution concepts and their interrelationships. Eur. J. Oper. Res. **41**, 86–100 (1989)
16. Fraser, N.M., Hipel, K.W.: Solving complex conflicts. IEEE Trans. Syst. Man Cybern. **9**, 805–816 (1979). https://doi.org/10.1109/TSMC.1979.4310131
17. Kuhn, J.R.D., Hipel, K.W., Fraser, N.M.: A coalition analysis algorithm with application to the Zimbabwe conflict. IEEE Trans. Syst. Man Cybern. **13**, 338–352 (1983). https://doi.org/10.1109/TSMC.1983.6313166
18. Kuhn, J.R.D., Hipel, K.W., Fraser, N.M.: Algorithm with application a coalition analysis conflict to the Zimbabwe. IEEE Trans. Syst. Man Cybern. **13**, 338–352 (1983)
19. Hipel, K.W., Meister, D.B.: Conflict analysis methodology for modelling coalitions in multilateral negotiations. Inf. Decis. Technol. Amsterdam **19**, 85–103 (1994)
20. Kilgour, D.M., Hipel, K.W., Peng, X.Y., Fang, L.P.: Coalition analysis in group decision support. Group Decis. Negot. **10**, 159–175 (2001)
21. Inohara, T., Hipel, K.W.: Coalition analysis in the graph model for conflict resolution. Syst. Eng. **11**, 343–359 (2008). https://doi.org/10.1002/sys.20104
22. Inohara, T., Hipel, K.W.: Interrelationships among non-cooperative and coalition stability concepts. J. Syst. Sci. Syst. Eng. **17**, 1–29 (2008). https://doi.org/10.1007/s11518-008-5070-1
23. Xu, H., Kilgour, D.M., Hipel, K.W.: Matrix representation and extension of coalition analysis in group decision support. Comput. Math. Appl. **60**, 1164–1176 (2010). https://doi.org/10.1016/j.camwa.2010.05.040
24. Xu, H., Hipel, K.W., Kilgour, D.M., Fang, L.: Conflict resolution using the graph model: strategic interactions in competition and cooperation. Stud. Syst. Decis. Control **153**, Springer,

Cham, Switzerland, https://doi.org/10.1007/978-3-319-77670-5, ISBN 978–3–319–77669–9 (hard copy), ISBN 978–3–319–77670–5 (eBook), 436 pp., 2018. (Chinese translation published by Science Press, Beijing, China in 2023, ISBN 978–03–071737–5)
25. Zhu, Z., Kilgour, D.M., Hipel, K.W.: A new approach to coalition analysis within the graph model. IEEE Trans. Syst. Man Cybern. Syst. (2018). https://doi.org/10.1109/TSMC.2018.2811402
26. Ge, B., Huang, Y., Hou, Z., Sun, J., You, Y., Yang, K.: Inverse approach to the graph model for conflict resolution under combinatorial behavior with two decision makers. In: 2023 18th Annual System of Systems Engineering Conference (SoSe), pp. 1–6 (2023)
27. Han, Y., Xu, H.Y., Fang, L.P., Hipel, K.W.: An integer programming approach to solving the inverse graph model for conflict resolution with two decision makers. Group Decis. Negot. **31**, 23–48 (2022). https://doi.org/10.1007/s10726-021-09755-w
28. Huang, Y., et al.: Solving the inverse graph model for conflict resolution using a hybrid metaheuristic algorithm. Eur. J. Oper. Res. **305**, 806–819 (2023). https://doi.org/10.1016/j.ejor.2022.06.052
29. Tao, L., Su, X., Javed, S.A.: Inverse preference optimization in the graph model for conflict resolution based on the genetic algorithm. Group Decis. Negot. **30**, 1085–1112 (2021). https://doi.org/10.1007/s10726-021-09748-9
30. Kinsara, R.A., Kilgour, D.M., Hipel, K.W.: Inverse approach to the graph model for conflict resolution. IEEE Trans. Syst. Man Cybern. Syst. **45**, 734–742 (2015). https://doi.org/10.1109/TSMC.2014.2376473
31. Hipel, K., Fang, L., Kilgour, D.: The graph model for conflict resolution: reflections on three decades of development. Group Decis. Negot. **29** (2020). https://doi.org/10.1007/s10726-019-09648-z
32. Zhu, Z., Kilgour, D.M., Hipel, K.W.: A new approach to coalition analysis within the graph model. IEEE Trans. Syst. Man Cybern. Syst. **50**, 2231–2241 (2020). https://doi.org/10.1109/TSMC.2018.2811402
33. Zhao, S., Xu, H., Hipel, K.W., Fang, L.: Mixed coalitional stabilities with full participation of sanctioning opponents within the graph model for conflict resolution. IEEE Trans. Syst. Man Cybern. Syst. **51**, 3911–3925 (2021). https://doi.org/10.1109/TSMC.2019.2950673
34. Zhu, Z., Kilgour, D.M., Hipel, K.W., Yu, J.: Interrelationships of Non-cooperative, Classical and Pareto coalitional stability definitions. Eur J Oper Res. **321**, 884–894 (2025). https://doi.org/10.1016/j.ejor.2024.10.035
35. Kilgour, D.M., Hipel, K.W.: The graph model for conflict resolution: past, present, and future. Group Decis. Negot. **14**, 441–460 (2005)
36. Inohara, T., Hipel, K.W.: Coalition analysis in the graph model for conflict resolution. Syst. Eng. **11**, 343–359 (2008). https://doi.org/10.1002/sys.20104

Triangular Fuzzy Preferences in the Graph Model with Two Decision-Makers

Tianyang Gu, Bingfeng Ge[✉], Wanying Wei, Sining Han, Zihui Liu, and Chi Wang

College of Systems Engineering, National University of Defense Technology, Changsha, China
{gty2wsl,bingfengge,weiwanying,wangchi0214}@nudt.edu.cn

Abstract. Decision-making in strategic conflicts often involves uncertain preferences. Triangular fuzzy preferences offer an intuitive and flexible representation of uncertainty. Therefore, this study aims to incorporate triangular fuzzy preferences into the graph model for conflict resolution (GMCR). A new framework for GMCR with two decision-makers (DMs) is developed, integrating triangular fuzzy preference relations to represent and analyze conflicts under uncertainty. Within the framework, four basic stability definitions are extended for two DMs to model with triangular fuzzy preferences. More specifically, triangular fuzzy Nash stability, triangular fuzzy general metarationality, triangular fuzzy symmetric metarationality and triangular fuzzy sequential stability provide a comprehensive description of human behavior patterns. A state is triangular fuzzy stable for a DM if there is no reward to leave the state, where reward is assessed according to triangular fuzzy relative strength of preference (TFRSP) and triangular fuzzy satisficing threshold (TFST). Furthermore, a triangular fuzzy equilibrium, which is triangular fuzzy stable for all DMs, represents a possible solution to the strategic conflict. Finally, an illustrative example is applied to demonstrate the applicability of the proposed framework in practice.

Keywords: Conflict Resolution · Triangular Fuzzy Preferences · GMCR · Stability Analysis

1 Introduction

Strategic conflicts are complex systems characterized by dynamic interactions among two or more decision-makes (DMs), where each DM independently makes strategic choices, collectively determining the final state of the system [1]. Strategic conflicts span a diverse array of domains, including politics, military affairs, economics and societal issues, reflecting their pervasive role throughout human history [1, 2]. From landmark events such as the Cuban Missile Crisis [3] to localized disputes like the Elmira groundwater contamination conflict [4], confrontations emerge due to divergences among DMs in interests, goals, values, or resource distribution [5].

To better understand and tackle strategic conflicts, several approaches have been developed within game theory [6]. However, classical game theory often proves unreliable and difficult to apply in numerous conflicts due to data limitations and single-step foresight. Based on the framework of classical game theory, the graph model for conflict resolution (GMCR) was extended from metagame analysis [7] and conflict analysis [8]. Through systemic modeling and analysis [2], GMCR integrates concepts from game theory, set theory, and graph theory to conflict resolution. GMCR requires only the relative preferences of DMs, facilitating a combination of both qualitative and quantitative analysis [1]. Thus, GMCR has been widely used in various real-world conflicts, including complex systems architecting [9], water resources allocation [10], arms control [11], and elsewhere.

GMCR systematically comprises two modules [1]: a modeling module, which establishes a conflict graph model within a mathematical framework, and an analysis module, which investigates the model through stability analysis and follow-up analyses [12]. In the modeling module, a key parameter of the graph model is the preference information. Due to the complexity and uncertainty, various preference structures have been developed, such as fuzzy preferences [5], grey preferences [13], unknown preferences [14] and belief preferences [15]. In the framework of fuzzy theory, Bashar et al. [5] developed a fuzzy preference structure for GMCR based on fuzzy relations, which was later extended to incorporate interval fuzzy preferences [16]. Wu et.al [17] proposed the incomplete reciprocal preference relation to accommodate both complete and incomplete fuzzy preferences. Similarly, Wang et al. [10] integrated intuitionistic fuzzy preferences into GMCR to represent the DMs' uncertain preferences. Wu et al. [18] introduced hesitant fuzzy preferences for conflict situations involving composite DMs. Furthermore, Yu et al. [19] put forward a hybrid preference structure covering the fuzzy and unknown preferences. However, the research on GMCR using triangular fuzzy preferences remains unexplored.

Due to the increasing complexity and uncertainty of strategic conflicts, DMs exhibit varying degrees of certainty during the analysis process. Among the different forms of uncertain preferences, triangular fuzzy preferences provide a distinct advantage by effectively depicting the range of preference variations while emphasizing the most probable preference [20]. For instance, a government might assess a policy outcome as "approximately acceptable," allowing for a degree of tolerance toward both more favorable and less favorable alternatives. Such inherent vagueness corresponds well with the structural characteristics of triangular fuzzy preferences. Thus, this study aims to extend the classical GMCR framework with two DMs by incorporating triangular fuzzy preferences, enabling a more flexible representation of preference uncertainty. The remainder of this paper is organized as follows. In Sect. 2, the classical GMCR and triangular fuzzy preferences are briefly reviewed. Section 3 introduces the triangular fuzzy framework for GMCR with two DMs. Section 4 presents an illustrative example to demonstrate the application and interpretation of the proposed framework. Finally, appropriate conclusions and future work are drawn in Sect. 5.

2 Preliminaries

2.1 Graph Model for Conflict Resolution

The resolution of conflicts within the GMCR framework encompasses a modeling stage and an analysis stage. The modeling stage focuses on analyzing the conflict's background and extracting model parameters, including DMs, available options, feasible states, state transitions and preference information [21]. The analysis stage is concerned with identifying equilibriums under different stability definitions and conducting follow-up analyses, such as status quo analysis [22], coalition analysis [23], and sensitivity analysis [24].

Definition 1: Formally, a graph model with two or more DMs is represented as a four-tuple structure $G = (N, S, \{A_k, P_k\}_{k \in N})$, where

1) $N = \{1, 2, ..., n\}$ represents the finite set of DMs with $n \geq 2$;
2) $S = \{s_1, s_2, ..., s_m\}$ represents the finite set of feasible states, and m denotes the number of feasible states;
3) $A_k \subseteq S \times S$ represents the finite set of oriented arcs containing the one-step movements controlled by DM k;
4) P_k represents the relative preference of DM k, represented by a pair of binary preference relations $\{\succ, \sim\}$, where $s_1 \succ_k s_2 (s_1, s_2 \in S)$ means DM k strictly prefers s_1 to s_2, and $s_1 \sim_k s_2$ indicates s_1 and s_2 are indifferent for DM k.

Definition 2: For $k \in N$ and $s \in S$, $R_k(s) = \{t \in S | (s, t) \in A_k\}$ denotes the unilateral movement (UM) list of DM k from state s.

Definition 3: For $k \in N$ and $s \in S$, $R_k^+(s) = \{t \in R_k(s) | t \succ_k s\}$ denotes the unilateral improvement (UI) list of DM k from state s.

Before moving, DMs carefully assess both their own and opponents' strategic options based on specific behavior patterns. This process is referred to as stability analysis. A conflict is considered resolved if and only if a resolution scheme is accepted by all involved DMs. The four stability definitions [2] for a graph model with n DMs, including Nash Stability, General Metarationality (GMR), Symmetric Metarationality (SMR), and Sequential Stability (SEQ).

2.2 Triangular Fuzzy Preference Relation

The fuzzy preference relation represents one of the earliest formal approaches to modeling uncertain preferences by using the fuzzy preference degree. However, the fuzzy preference degree is typically represented for by a crisp number. To capture more uncertainty in preference modeling, the concept of the triangular fuzzy preference relation is introduced [25]. In this approach, the preference for one state over another is represented by a triangular fuzzy number. A triangular fuzzy number is characterized by a triplet (l, m, u), where l denotes the lower bound, m denotes the most probable value, and u denotes the upper bound [26]. This representation captures the range of possible preferences, reflecting varying degrees of uncertainty. A formal definition is provided next.

Definition 4: Let $S = \{s_1, s_2, ..., s_m\}$ denote the set of states. A triangular fuzzy preference relation on S is represented by a triangular fuzzy judgment matrix $\tilde{R} = (\tilde{r}_{ij})_{m \times m}$, where $\tilde{r}_{ij} = (l_{ij}, m_{ij}, u_{ij})$ illustrates the triangular fuzzy preference degree of state s_i to

s_j. For all $i, j = 1, 2, ..., m$, l_{ij}, m_{ij} and u_{ij} are assumed to satisfy $0 \leq l_{ij}, m_{ij}, u_{ij} \leq 1$, $l_{ij} + u_{ji} = m_{ij} + m_{ji} = u_{ij} + l_{ji} = 1$, $l_{ii} = m_{ii} = u_{ii} = 0.5$.

3 A Triangular Fuzzy Framework for the Graph Model with Two Decision-Makers

In this section, a triangular fuzzy preference structure is integrated into the framework of GMCR. The concepts of a DM's triangular fuzzy relative strength of preference (TFRSP), defuzzification of triangular fuzzy relative strength of preference (DTFRSP) and triangular fuzzy satisficing threshold (TFST) are firstly introduced. Subsequently, the triangular fuzzy stability definitions are proposed to model and analyze the conflict problems with two DMs.

3.1 Basic Concepts for the Triangular Fuzzy Framework

A DM's triangular fuzzy preference over feasible states is a pairwise relation that quantifies the degree to which one state is preferred over another. According to Definition 9, the triangular fuzzy preference is characterized by three preference degrees. The degree l_{ij} represents the lower bound of the preference. The degree m_{ij} reflects the most probable preference, typically regarded as the optimal choice. The degree u_{ij} corresponds to the upper bound of the preference. Due to additive reciprocity, the number $\tilde{r}_{ji} = (1 - u_{ij}, 1 - m_{ij}, 1 - l_{ij})$ can be interpreted as the degree to which state s_i is not preferred over s_j. Hence, the following definition describes the intensity of preference for a state relative to another, which will be referred to as a DM's TFRSP.

Definition 5: Let $k \in N$, and let $\tilde{r}^k(s_i, s_j) = \tilde{r}_{ij}^k$ denote the triangular fuzzy preference degree of state s_i over s_j for DM k. Then, DM k's TFRSP for state s_i over s_j is denoted as $\tilde{\alpha}^k(s_i, s_j) = \tilde{r}^k(s_i, s_j) - \tilde{r}^k(s_j, s_i) = (l_{ij} - u_{ji}, m_{ij} - m_{ji}, u_{ij} - l_{ji})$.

Obviously, for all $i, j = 1, 2, ..., m$, $-1 \leq l_{ij} - u_{ji}, m_{ij} - m_{ji}, u_{ij} - l_{ji} \leq 1$. In particular:

(1) $\tilde{\alpha}^k(s_i, s_j) = (1, 1, 1)$ indicates that DM k definitely prefers state s_i over s_j;
(2) $\tilde{\alpha}^k(s_i, s_j) = (0, 0, 0)$ means that DM k is equally likely to prefer state s_i over s_j;
(3) $\tilde{\alpha}^k(s_i, s_j) = (-1, -1, -1)$ indicates that DM k definitely prefer state s_j over s_i.

The number $\tilde{\alpha}^k(s_i, s_j)$ measures the range and center of the relative certainty of DM k's triangular fuzzy preference for state s_i over s_j. To facilitate subsequent computations and analyses, it is essential to clarify the TFRSP. The center of gravity (COG) defuzzification technique [27] is utilized to defuzzify the TFRSP, thereby generating the DTFRSP. The formal definition of DTFRSP is given as follows.

Definition 6: Let $k \in N$. The DM k's DTFRSP for state s_i over s_j is denoted as $\overline{\alpha}^k(s_i, s_j) = \frac{l_{ij} + m_{ij} + u_{ij} - l_{ji} - m_{ji} - u_{ji}}{3}$, which is the COG of $\tilde{\alpha}^k(s_i, s_j)$. Denoting $\overline{\alpha}^k(s_i, s_j) =$

$\overline{\alpha}_{ij}^k$ for any $i, j = 1, 2, ..., m$, the DM k's DTFRSP over S can be represented by the matrix $\left(\overline{\alpha}_{ij}^k\right)_{m \times m}$.

In analyzing a graph model, the key work is to assess whether a DM should remain in a focal state or transition to another state. Every DM in the model is capable of selecting a level of DTFRSP to determine whether such a move is worthwhile. This level of DTFRSP is referred to as a DM's TFST. More formally, the following definition is given.

Definition 7: Let $k \in N$. DM k would be willing to move from state s to s_i if and only if $\overline{\alpha}^k(s_i, s) \geq \lambda_k$, where λ_k is called the TFST of DM k. Note that for all DMs, $0 < \lambda_k \leq 1$.

Based on the given TFST, the set of preferred states for DM k can be determined. Thus, a new definition of triangular fuzzy unilateral improvement (TFUI) is introduced, which serves to describe the preferred states relative to a give state s.

Definition 8: Let $s \in S$ and $k \in N$. A state $s_i \in R_k(s)$ is called a TFUI from state s for DM k if and only if $\overline{\alpha}^k(s_i, s) \geq \lambda_k$.

According to Definition 2, $R_k(s)$ is used to represents the set of reachable states by UM from the initial state s for DM k. Thus, the definition of triangular fuzzy unilateral improvement list (TFUIL) is given to represent all the TFUI states from state s in $R_k(s)$.

Definition 9: Let $s \in S$ and $k \in N$. The TFUIL from state s for DM k is denoted as $\tilde{R}_{k,\lambda_k}^+(s) = \left\{s_i \in R_k(s) | \overline{\alpha}^k(s_i, s) \geq \lambda_k\right\}$. For simplicity, let $\tilde{R}_{k,\lambda_k}^+(s)$ be denoted as $\tilde{R}_k^+(s)$.

3.2 Triangular Fuzzy Stabilities for the Graph Model with Two Decision-Makers

The stability analysis is the key analysis procedure in a graph model, which is used to identify the stability of a given state for a DM. To carry out stability analysis in the graph model with two DMs, new stability definitions are needed. Thus, four basic stability definitions under classic GMCR are extended to triangular fuzzy stability definitions based on the triangular fuzzy preferences and behavior patterns. For simplicity, the set $N = \{k, l\}$ is assumed to represent two different DMs in the graph model with two DMs. Accordingly, the TFSTs are denoted as λ_k and λ_l.

Definition 10: Let $s \in S$ and $k \in N$. The state s is triangular fuzzy Nash stability (TFNash) for DM k if and only if $\tilde{R}_k^+(s) = \emptyset$.

Under TFNash, a DM just considers his or her potential TFUIs when deciding whether to move from a given state and ignores other DMs' possible countermoves. Thus, state s is TFNash for DM k if and only if DM k has no TFUIs from state s.

Definition 11: Let $s \in S$ and $k \in N$. The state s is triangular fuzzy general metarationality (TFGMR) for DM k if and only for every $s_1 \in \tilde{R}_k^+(s)$, there exists $s_2 \in R_l(s_1)$ with $\overline{\alpha}^k(s_2, s) < \lambda_k$.

Under TFGMR, a DM assesses whether each of his or her potential TFUIs will be sanctioned by one of the opponent's UMs. Note that the focal DM does not consider whether the opponent would benefit from this sanction. Furthermore, if a DM has no TFUIs from a state s, then s is also TFGMR.

Definition 12: Let $s \in S$ and $k \in N$. The state s is triangular fuzzy symmetric metarationality (TFSMR) for DM k if and only if for every $s_1 \in \tilde{R}_k^+(s)$, there exists $s_2 \in R_l(s_1)$ such that $\overline{\alpha}^k(s_2, s) < \lambda_k$ and $\overline{\alpha}^k(s_3, s) < \lambda_k$ for all $s_3 \in R_k(s_2)$.

Under TFSMR, a DM considers one more step in comparison with TFGMR. When a sanction is imposed by the opponent, the focal DM seeks to determine whether the sanction can be evaded by using a UM. Thus, if the focal DM cannot escape the sanction of the opponent, the given state is TFSMR for the DM. Furthermore, if a DM has no TFUIs from a state s, then s is also TFSMR.

Definition 13: Let $s \in S$ and $k \in N$. The state s is triangular fuzzy sequential stability (TFSEQ) for DM k if and only for every $s_1 \in \tilde{R}_k^+(s)$, there exists $s_2 \in \tilde{R}_l^+(s_1)$ with $\overline{\alpha}^k(s_2, s) < \lambda_k$.

The stability definition of TFSEQ is similar to the stability definition of TFGMR. The key difference under TFSEQ is that the focal DM just considers the credible sanctions by the opponent when considering his or her potential TFUIs. Thus, TFSEQ depends not only on the focal DM's TFST but on the opponent's TFST. Furthermore, if a DM has no TFUIs from a state s, then s is also TFSEQ.

Definition 14: In the triangular fuzzy framework for the graph model with two DMs, a state $s \in S$ is referred to as a triangular fuzzy equilibrium (TFE) if and only if it is triangular fuzzy stable for every DM under specific stability definition.

4 An Illustrative Example

This section illustrates the triangular fuzzy framework for GMCR with two DMs by analyzing the Cuban Missile Crisis. Firstly, the background of the Cuban Missile Crisis is introduced [28]. Next, this conflict is systematically modeled using the proposed framework. Finally, the stability and status quo analysis are conducted to identify the desired equilibrium and the evolution of the conflict.

4.1 Background

The Cuban Missile Crisis, which occurred in October 1962, involved two primary DMs: the United States (US) and the Union of Soviet Socialist Republics (USSR). This confrontation was triggered when the US discovered that the USSR had secretly deployed missiles in Cuba. Subsequently, the conflict rapidly escalated as the US imposed severe countermoves, including a naval blockade of Cuba. Ultimately, the conflict deescalated with the withdrawal of USSR's missiles from Cuba. Widely regarded as one of the most critical moments of the Cold War, the conflict brought the world to the brink of a nuclear war. Thus, the study and understanding of this conflict remain crucial for effectively managing current and future international conflicts.

4.2 Conflict Modeling

DMs and Their Options: To model this conflict, it is essential to first identify the DMs. In this context, the primary DMs are US and USSR, each of which possesses distinct strategic options. US controls two strategic options: 1) launch an airstrike on the Cuban missile base (Airstrike); and 2) impose a naval blockade on Cuba (Blockade). In response to US's actions, USSR also controls two strategic options: 1) withdraw the missiles deployed in Cuba (Withdrawal); and 2) escalate the conflict against US (Escalation).

Feasible States and State Transitions: In the conflict model, there are two DMs and four strategic options. Each state is represented as a combination of all the options in form of "Y/N", thus there are theoretically $2^4 = 16$ states. However, some infeasible states need to be eliminated. After applying the state reduction rules, which include the mutual exclusion rule, the at least one choice rule, the strategy dependency rule and the direct removal rule, 12 feasible states remain, as shown in Table 1. In the table, "Y" indicates that an option is selected and "N" indicates that an option is not selected. For example, s_1 represents a situation where no options are selected, while s_2 indicates that US has chosen to conduct an airstrike.

After determining the feasible states, the state transition diagrams for US and USSR are respectively presented in Fig. 1 and Fig. 2. The points labeled with corresponding numbers in the diagrams represent the 12 feasible states. The directed edges illustrate the transition from one state to another under the option of the focal DM.

Table 1. Feasible states of the conflict.

DMs	Options	s_1	s_2	s_3	s_4	s_5	s_6	s_7	s_8	s_9	s_{10}	s_{11}	s_{12}
US	Airstrike	N	Y	N	Y	N	Y	N	Y	N	Y	N	Y
	Blockade	N	N	Y	Y	N	N	Y	Y	N	N	Y	Y
USSR	Withdrawal	N	N	N	N	Y	Y	Y	Y	N	N	N	N
	Escalation	N	N	N	N	N	N	N	N	Y	Y	Y	Y

Preference Information: Both DMs possess triangular fuzzy preferences, which means their preference values are expressed within a defined range, with each value encompassing a most probable estimate. Table 2 presents the triangular fuzzy preference matrices for US and USSR, denoted by $\tilde{R}^{US}$ and $\tilde{R}^{USSR}$, respectively. For example, in the $\tilde{R}^{US}$, the value in the second row and fourth column represents the degree of preference US assigns to s_2 over s_4.

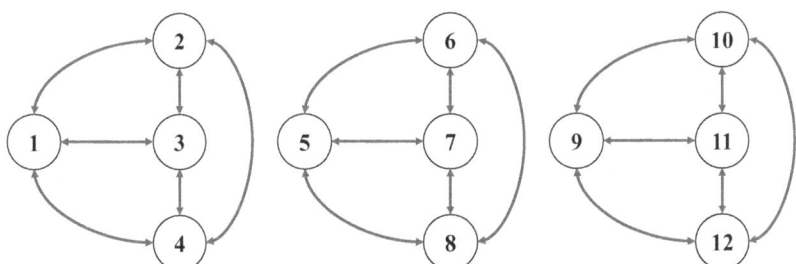

Fig. 1. The state transitions of US.

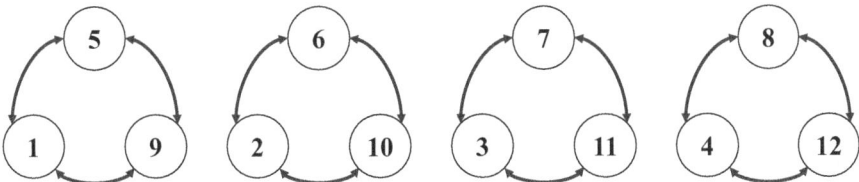

Fig. 2. The state transitions of USSR.

4.3 Stability Analysis

The results of the stability analysis based on triangular fuzzy preferences are derived using triangular fuzzy stability definitions for two DMs. First, a group of TFSTs for US and USSR is determined, with parameter values of $\lambda_{US} = 0.2$ and $\lambda_{USSR} = 0.1$. Following this, the triangular fuzzy stability definitions, including Definitions 15–18, are applied to identify the stable states and equilibrium. The results of the stability analysis are displayed in Table 3.

If a state is an equilibrium under all stability definitions, the state is called the strongest equilibrium. According to Table 3, it can be observed that there is no strongest equilibrium in the conflict, indicating that the conflict is characterized by high uncertainty and dynamic complexity. This requires a thorough analysis of the conflict situation and strategic decision-making. States s_5 and s_7 satisfy the equilibrium conditions under the definitions of TFGMR, TFSMR, and TFSEQ, making them relatively strong equilibrium.

In state s_5, US takes no action, while USSR chooses to withdraw the missiles deployed in Cuba. However, achieving s_5 would require USSR to unilaterally degrade

Table 2. The triangular fuzzy preference matrices for US and USSR.

$\tilde{R}^{US} =$

	s_1	s_2	s_3	s_4	s_5	s_6	s_7	s_8	s_9	s_{10}	s_{11}	s_{12}
s_1	(0.5, 0.5, 0.5)	(0.2, 0.3, 0.4)	(0.1, 0.2, 0.3)	(0.3, 0.4, 0.5)	(0.0, 0.0, 0.0)	(0.0, 0.0, 0.0)	(0.0, 0.0, 0.0)	(0.0, 0.0, 0.0)	(1.0, 1.0, 1.0)	(0.8, 0.9, 1.0)	(1.0, 1.0, 1.0)	(0.8, 0.9, 1.0)
s_2	(0.6, 0.7, 0.8)	(0.5, 0.5, 0.5)	(0.3, 0.4, 0.5)	(0.6, 0.8, 0.9)	(0.0, 0.0, 0.0)	(0.0, 0.0, 0.0)	(0.0, 0.0, 0.0)	(0.0, 0.0, 0.0)	(0.1, 0.2, 0.3)	(1.0, 1.0, 1.0)	(0.8, 0.9, 1.0)	(1.0, 1.0, 1.0) (0.8, 0.9, 1.0)
s_3	(0.7, 0.8, 0.9)	(0.5, 0.6, 0.7)	(0.5, 0.5, 0.5)	(0.7, 0.8, 0.9)	(0.0, 0.0, 0.0)	(0.1, 0.2, 0.3)	(0.0, 0.0, 0.0)	(0.1, 0.3, 0.4)	(1.0, 1.0, 1.0)	(1.0, 1.0, 1.0)	(1.0, 1.0, 1.0)	(1.0, 1.0, 1.0)
s_4	(0.5, 0.6, 0.7)	(0.1, 0.2, 0.4)	(0.1, 0.2, 0.3)	(0.5, 0.5, 0.5)	(0.0, 0.0, 0.0)	(0.0, 0.0, 0.0)	(0.0, 0.0, 0.0)	(0.2, 0.3)	(0.0, 0.1, 0.2)	(1.0, 1.0, 1.0)	(0.8, 0.9, 1.0)	(1.0, 1.0, 1.0) (0.7, 0.9, 1.0)
s_5	(1.0, 1.0, 1.0)	(1.0, 1.0, 1.0)	(1.0, 1.0, 1.0)	(1.0, 1.0, 1.0)	(0.5, 0.5, 0.5)	(0.6, 0.7, 0.8)	(0.5, 0.6, 0.7)	(0.7, 0.8, 0.9)	(1.0, 1.0, 1.0)	(1.0, 1.0, 1.0)	(1.0, 1.0, 1.0)	(1.0, 1.0, 1.0)
s_6	(1.0, 1.0, 1.0)	(1.0, 1.0, 1.0)	(0.7, 0.8, 0.9)	(1.0, 1.0, 1.0)	(0.2, 0.3, 0.4)	(0.5, 0.5, 0.5)	(0.3, 0.4, 0.5)	(0.6, 0.7, 0.8)	(1.0, 1.0, 1.0)	(1.0, 1.0, 1.0)	(1.0, 1.0, 1.0)	(1.0, 1.0, 1.0)
s_7	(1.0, 1.0, 1.0)	(1.0, 1.0, 1.0)	(1.0, 1.0, 1.0)	(0.7, 0.8, 1.0)	(0.3, 0.4, 0.5)	(0.5, 0.6, 0.7)	(0.5, 0.5, 0.5)	(0.7, 0.8, 0.9)	(1.0, 1.0, 1.0)	(1.0, 1.0, 1.0)	(1.0, 1.0, 1.0)	(1.0, 1.0, 1.0)
s_8	(1.0, 1.0, 1.0)	(0.7, 0.8, 0.9)	(0.6, 0.7, 0.9)	(0.8, 0.9, 1.0)	(0.1, 0.2, 0.3)	(0.2, 0.3, 0.4)	(0.1, 0.2, 0.3)	(0.5, 0.5, 0.5)	(1.0, 1.0, 1.0)	(1.0, 1.0, 1.0)	(1.0, 1.0, 1.0)	(0.8, 0.9, 1.0)
s_9	(0.0, 0.0, 0.0)	(0.0, 0.0, 0.0)	(0.0, 0.0, 0.0)	(0.0, 0.0, 0.0)	(0.0, 0.0, 0.0)	(0.0, 0.0, 0.0)	(0.0, 0.0, 0.0)	(0.0, 0.0, 0.0)	(0.5, 0.5, 0.5)	(0.1, 0.2, 0.3)	(0.2, 0.3, 0.4)	(0.0, 0.1, 0.2)
s_{10}	(0.0, 0.1, 0.2)	(0.0, 0.1, 0.2)	(0.0, 0.0, 0.0)	(0.0, 0.1, 0.2)	(0.0, 0.0, 0.0)	(0.0, 0.0, 0.0)	(0.0, 0.0, 0.0)	(0.0, 0.0, 0.0)	(0.7, 0.8, 0.9)	(0.5, 0.5, 0.5)	(0.5, 0.6, 0.7)	(0.3, 0.4, 0.5)
s_{11}	(0.0, 0.0, 0.0)	(0.0, 0.0, 0.0)	(0.0, 0.0, 0.0)	(0.0, 0.0, 0.0)	(0.0, 0.0, 0.0)	(0.0, 0.0, 0.0)	(0.0, 0.0, 0.0)	(0.0, 0.0, 0.0)	(0.6, 0.7, 0.8)	(0.3, 0.4, 0.5)	(0.5, 0.5, 0.5)	(0.1, 0.2, 0.3)
s_{12}	(0.0, 0.1, 0.2)	(0.0, 0.1, 0.2)	(0.0, 0.0, 0.0)	(0.0, 0.1, 0.3)	(0.0, 0.0, 0.0)	(0.0, 0.0, 0.0)	(0.0, 0.0, 0.0)	(0.0, 0.1, 0.2)	(0.8, 0.9, 1.0)	(0.5, 0.6, 0.7)	(0.7, 0.8, 0.9)	(0.5, 0.5, 0.5)

$\tilde{R}^{USSR} =$

	s_1	s_2	s_3	s_4	s_5	s_6	s_7	s_8	s_9	s_{10}	s_{11}	s_{12}
s_1	(0.5, 0.5, 0.5)	(1.0, 1.0, 1.0)	(0.8, 0.9, 1.0)	(1.0, 1.0, 1.0)	(0.6, 0.7, 0.8)	(1.0, 1.0, 1.0)	(0.7, 0.8, 0.9)	(1.0, 1.0, 1.0)	(1.0, 1.0, 1.0)	(1.0, 1.0, 1.0)	(1.0, 1.0, 1.0)	(1.0, 1.0, 1.0)
s_2	(0.0, 0.0, 0.0)	(0.5, 0.5, 0.5)	(0.2, 0.3, 0.4)	(0.6, 0.7, 0.8)	(0.1, 0.2, 0.3)	(0.3, 0.4, 0.5)	(0.1, 0.2, 0.3)	(0.5, 0.6, 0.7)	(1.0, 1.0, 1.0)	(1.0, 1.0, 1.0)	(1.0, 1.0, 1.0)	(1.0, 1.0, 1.0)
s_3	(0.0, 0.1, 0.2)	(0.6, 0.7, 0.8)	(0.5, 0.5, 0.5)	(0.7, 0.8, 0.9)	(0.3, 0.4, 0.5)	(0.5, 0.6, 0.7)	(0.3, 0.4, 0.5)	(0.6, 0.7, 0.8)	(1.0, 1.0, 1.0)	(1.0, 1.0, 1.0)	(0.8, 0.9, 1.0)	(1.0, 1.0, 1.0)
s_4	(0.0, 0.0, 0.0)	(0.2, 0.3, 0.4)	(0.1, 0.2, 0.3)	(0.5, 0.5, 0.5)	(0.0, 0.1, 0.2)	(0.1, 0.2, 0.3)	(0.0, 0.1, 0.2)	(0.4, 0.4, 0.5)	(1.0, 1.0, 1.0)	(0.7, 0.8, 0.9)	(1.0, 1.0, 1.0)	(0.6, 0.7, 0.8)
s_5	(0.2, 0.3, 0.4)	(0.7, 0.8, 0.9)	(0.5, 0.6, 0.7)	(0.8, 0.9, 1.0)	(0.5, 0.5, 0.5)	(0.6, 0.7, 0.8)	(0.5, 0.6, 0.7)	(0.7, 0.8, 0.9)	(1.0, 1.0, 1.0)	(1.0, 1.0, 1.0)	(1.0, 1.0, 1.0)	(1.0, 1.0, 1.0)
s_6	(0.0, 0.0, 0.0)	(0.5, 0.6, 0.7)	(0.2, 0.3, 0.4)	(0.7, 0.8, 0.9)	(0.2, 0.3, 0.4)	(0.5, 0.5, 0.5)	(0.1, 0.2, 0.3)	(0.7, 0.8, 0.9)	(1.0, 1.0, 1.0)	(0.8, 0.9, 1.0)	(1.0, 1.0, 1.0)	(0.8, 0.9, 1.0)
s_7	(0.1, 0.2, 0.3)	(0.7, 0.8, 0.9)	(0.5, 0.6, 0.7)	(0.8, 0.9, 1.0)	(0.3, 0.4, 0.5)	(0.7, 0.8, 0.9)	(0.5, 0.5, 0.5)	(0.8, 0.9, 1.0)	(1.0, 1.0, 1.0)	(1.0, 1.0, 1.0)	(1.0, 1.0, 1.0)	(0.8, 0.9, 1.0)
s_8	(0.0, 0.0, 0.0)	(0.3, 0.4, 0.5)	(0.2, 0.3, 0.4)	(0.5, 0.6, 0.6)	(0.1, 0.2, 0.3)	(0.1, 0.2, 0.3)	(0.0, 0.1, 0.2)	(0.5, 0.5, 0.5)	(1.0, 1.0, 1.0)	(1.0, 1.0, 1.0)	(1.0, 1.0, 1.0)	(0.7, 0.8, 0.9)
s_9	(0.0, 0.0, 0.0)	(0.0, 0.0, 0.0)	(0.0, 0.0, 0.0)	(0.0, 0.0, 0.0)	(0.0, 0.0, 0.0)	(0.0, 0.0, 0.0)	(0.0, 0.0, 0.0)	(0.0, 0.0, 0.0)	(0.5, 0.5, 0.5)	(0.1, 0.2, 0.3)	(0.2, 0.3, 0.4)	(0.0, 0.1, 0.2)
s_{10}	(0.0, 0.0, 0.0)	(0.0, 0.1, 0.2)	(0.0, 0.0, 0.0)	(0.1, 0.2, 0.3)	(0.0, 0.0, 0.0)	(0.0, 0.1, 0.2)	(0.0, 0.0, 0.0)	(0.0, 0.0, 0.0)	(0.7, 0.8, 0.9)	(0.5, 0.5, 0.5)	(0.8, 0.9, 1.0)	(0.3, 0.4, 0.5)
s_{11}	(0.0, 0.0, 0.0)	(0.0, 0.0, 0.0)	(0.0, 0.1, 0.2)	(0.0, 0.1, 0.2)	(0.0, 0.0, 0.0)	(0.0, 0.0, 0.0)	(0.0, 0.0, 0.0)	(0.0, 0.0, 0.0)	(0.6, 0.7, 0.8)	(0.0, 0.1, 0.2)	(0.5, 0.5, 0.5)	(0.2, 0.3, 0.4)
s_{12}	(0.0, 0.0, 0.0)	(0.0, 0.1, 0.2)	(0.0, 0.0, 0.0)	(0.2, 0.3, 0.4)	(0.0, 0.0, 0.0)	(0.0, 0.1, 0.2)	(0.0, 0.1, 0.2)	(0.1, 0.2, 0.3)	(0.8, 0.9, 1.0)	(0.5, 0.6, 0.7)	(0.6, 0.7, 0.8)	(0.5, 0.5, 0.5)

from s_1 (the status quo), which is evidently unacceptable. Therefore, this state is unlikely to serve as a solution to the conflict.

In state s_7, US imposes a naval blockade, and USSR agrees to removes the missiles from Cuba. This represents a relatively appropriate compromise: US refrains from conducting airstrikes, and USSR avoids escalating the conflict. Notably, from USSR's perspective, there are no TFUIs from s_7, making it TFNash for USSR. Thus, s_7 emerges as the most possible solution to the conflict, aligning with the historical outcome.

4.4 Status Quo Analysis

As one of the useful analysis tools in the graph model, status quo analysis is used to address the dynamic aspects of conflict and provides valuable insights into its temporal evolution.

Figure 3 illustrates the transition from the initial state s_1 to the equilibrium s_7 and the subsequent return to state s_1. In the diagram, red directed arcs represent unilateral moves controlled by US, while blue directed arcs indicate unilateral moves controlled by USSR.

In state s_1, neither side selected any strategic options. Subsequently, US discovered USSR's missiles deployed in Cuba and announced a military blockade of the island, escalating the situation to state s_3. Following intense political negotiations, USSR declared its intention to withdraw the missiles from Cuba, bringing the conflict to the equilibrium s_7, where a mutual compromise was reached. This phase marked the resolution of the conflict through reciprocal concessions. Subsequently, due to USSR's withdrawal of missiles, US lifted the naval blockade of Cuba, shifting the conflict to state s_5. Finally, with the full resolution of the crisis, the conflict returned to state s_1, thereby completing the cycle.

Triangular Fuzzy Preferences in the Graph Model   133

Table 3. Stability results of the conflict.

States	TFNash			TFGMR			TFSMR			TFSEQ		
	US	USSR	TFE	US	USSR	TFE	US	USSR	TFE	US	USSR	TFE
s_1		✓		✓	✓	✓	✓	✓	✓		✓	
s_2				✓	✓	✓	✓	✓	✓			
s_3	✓			✓	✓	✓	✓	✓	✓	✓		
s_4		✓		✓	✓	✓	✓	✓	✓		✓	
s_5	✓			✓	✓	✓	✓	✓	✓	✓	✓	✓
s_6		✓		✓	✓	✓	✓	✓	✓		✓	
s_7		✓		✓	✓	✓	✓	✓	✓	✓	✓	✓
s_8		✓		✓	✓	✓	✓	✓	✓		✓	
s_9												
s_{10}												
s_{11}												
s_{12}	✓			✓			✓			✓		

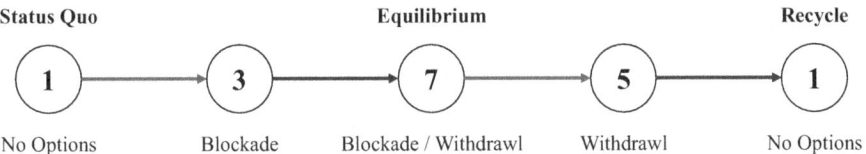

Fig. 3. The evolution path from the status quo.

5 Conclusion

In this paper, a triangular fuzzy preference framework has been developed for GMCR with two DMs. The framework incorporates preference uncertainty into decision making of conflict resolution in a flexible manner. Definitions for a DM's TFRSP, DTFRSP, TFST and TFUI are introduced, respectively. These definitions make it possible for triangular fuzzy preferences to be integrated in a graph model. Subsequently, four basic graph model stability definitions with two DMs are redefined as TFNash, TFGMR, TFSMR, and TFSEQ, respectively. Finally, by modeling and analyzing the Cuban Missile Crisis, the stability results and evolution path align closely with historical realities, thereby validating the effectiveness of the framework proposed.

This study systematically extends the classical GMCR with triangular fuzzy preferences. The proposed framework is a foundation to deal with conflicts involving more complexity and uncertainty. However, the framework is currently limited to two DMs, and the stability definitions are presented only in a logical form. Future work will aim to extend the framework to accommodate multiple DMs and develop a corresponding matrix representation to facilitate computational implementation.

Acknowledgement. This work was supported in part by the National Natural Science Foundation of China under Grants (71971213, and 72301288), and by the General Project of Postgraduate Scientific Research Innovation Project of Hunan Province (Grant No. CX20230086).

References

1. Xu, H., Hipel, K.W., Kilgour, D.M., et al.: Conflict resolution using the graph model: strategic interactions in competition and cooperation. Springer, Cham (2018)
2. Hipel, K.W., Fang, L., Kilgour, D.M.: The graph model for conflict resolution: reflections on three decades of development. Group Decis. Negot. **29**, 11–60 (2020)
3. Shaabani, S., Gordji, M.E.: Game theory and a new insight into how the Cuban missile crisis was resolved. Int. J. Cuban Stud. **15**(1), 39–49 (2023)
4. Wang, D., Huang, J., Xu, Y.: Matrix representations of the inverse problem in the graph model for conflict resolution with fuzzy preference. Appl. Soft Comput. **147**, 110786 (2023)
5. Bashar, M.A., Kilgour, D.M., Hipel, K.W.: Fuzzy preferences in the graph model for conflict resolution. IEEE Trans. Fuzzy Syst. **20**(4), 760–770 (2012)
6. Von Neumann, J., Morgenstern, O.: Theory of Games and Economic Behavior. Princeton University Press, Princeton (1944)
7. Howard, N.: Paradoxes of Rationality: Theory of Metagames and Political Behavior. MIT Press, Cambridge (1971)
8. Fraser, N.M., Hipel, K.W.: Conflict Analysis: Models and Resolutions. North-Holland, New York (1984)
9. Ge, B., Hipel, K.W., Fang, L., et al.: An interactive portfolio decision analysis approach for system-of-systems architecting using the graph model for conflict resolution. IEEE Trans. Syst. Man Cybern. Syst. **44**(10), 1328–1346 (2014)
10. Wang, D., Huang, J., Xu, Y.: Integrating intuitionistic preferences into the graph model for conflict resolution with applications to an ecological compensation conflict in Taihu Lake basin. Appl. Soft Comput. **135**, 110036 (2023)
11. Langenegger, T.W., Hipel, K.W.: The strategy of escalation and negotiation: the Iran nuclear dispute. J. Syst. Sci. Syst. Eng. **28**, 434–448 (2019)
12. Wu, N., Xu, Y., Wang, H., et al.: Matrix representation and behavioral analysis in a graph model for conflict resolution with incomplete fuzzy preferences. IEEE Trans. Syst. Man Cybern. Syst. **54**(1), 300–311 (2023)
13. Kuang, H., Bashar, M.A., Hipel, K.W., et al.: Grey-based preference in a graph model for conflict resolution with multiple decision makers. IEEE Trans. Syst. Man Cybern. Syst. **45**(9), 1254–1267 (2015)
14. Li, K.W., Hipel, K.W., Kilgour, D.M., et al.: Preference uncertainty in the graph model for conflict resolution. IEEE Trans. Syst. Man Cybern.-Part A Syst. Hum. **34**(4), 507–520 (2004)
15. Huang, Y., Ge, B., Sun, J., et al.: Belief-based preference structure and elicitation in the graph model for conflict resolution. IEEE Trans. Syst. Man Cybern. Syst. **53**(2), 727–740 (2022)
16. Bashar, M.A., Obeidi, A., Kilgour, D.M., et al.: Modeling fuzzy and interval fuzzy preferences within a graph model framework. IEEE Trans. Fuzzy Syst. **24**(4), 765–778 (2016)
17. Wu, N., Xu, Y., Hipel, K.W.: The graph model for conflict resolution with incomplete fuzzy reciprocal preference relations. Fuzzy Sets Syst. **377**, 52–70 (2019)
18. Wu, N., Xu, Y., Kilgour, D.M., et al.: Composite decision makers in the graph model for conflict resolution: hesitant fuzzy preference modeling. IEEE Trans. Syst. Man Cybern. Syst. **51**(12), 7889–7902 (2021)
19. Yu, J., Hipel, K.W., Kilgour, D.M., et al.: Graph model under unknown and fuzzy preferences. IEEE Trans. Fuzzy Syst. **28**(2), 308–320 (2019)

20. Wang, F., Wan, S.: Additive consistent triangular fuzzy preference relation and likelihood comparison algorithm based group decision making. Appl. Intell. **53**(10), 12098–12113 (2023)
21. Huang, Y., Ge, B., Hou, Z., et al.: Multi-unmanned aerial vehicle cooperative air combat gamingbased on graph model for conflict resolution. Syst. Eng. Theory Pract. **43**(9), 2714–2725 (2023)
22. Li, K.W., Kilgour, D.M., Hipel, K.W.: Status quo analysis in the graph model for conflict resolution. J. Oper. Res. Soc. **56**(6), 699–707 (2005)
23. Inohara, T., Hipel, K.W.: Coalition analysis in the graph model for conflict resolution. Syst. Eng. **11**(4), 343–359 (2008)
24. Li, X., Sun, Y., Zhou, S., et al.: Grey preference for analyzing the influence of externality within the graph model for conflict resolution. Expert Syst. Appl. **249**, 123736 (2024)
25. Xu, Z.: A method for priorities of triangular fuzzy number complementary judgment matrices. Fuzzy Syst. Math. **16**(1), 47–50 (2002)
26. Wang, F.: Preference degree of triangular fuzzy numbers and its application to multi-attribute group decision making. Expert Syst. Appl.pl. **178**, 114982 (2021)
27. Voskoglou, M.G.: Defuzzification of fuzzy numbers for student assessment. Am. J. Appl. Math. Stat. **3**(5), 206–210 (2015)
28. Scott, L., Hughes, R.G.: The Cuban Missile Crisis: A Critical Reappraisal. Routledge, New York (2015)

Strength of Preference and Probabilistic Preference in the Graph Model for Conflict Resolution for Two Decision-Makers

Elton César dos Santos Silva[1](✉), Danielle Costa Morais[1], and Liping Fang[2]

[1] Graduate Program in Management Engineering, Universidade Federal de Pernambuco – UFPE, Avenida da Arquitetura, Cidade Universitária, Recife, PE 50740-550, Brazil
elton-ceesar@hotmail.com, dcmorais@insid.org.br

[2] Department of Mechanical, Industrial, and Mechatronics Engineering, Toronto Metropolitan University, 350 Victoria Street, Toronto, ON M5B 2K3, Canada
lfang@torontomu.ca

Abstract. This paper proposes a new hybrid preference structure combining strength of preference and probabilistic preference within the Graph Model for Conflict Resolution (GMCR) for two Decision-Makers (DMs). This novel preference structure allows a DM to strongly or mildly prefer one state over another, or to express their preferences probabilistically. Four stability definitions α-Nash, α, β-General Metarationality, α, β-Symmetric Metarationality, and α, β, γ-Sequential Stability are extended to include general, strong and weak stabilities for conflicts with two DMs. The new preference structure, a combination of three-level strength of preference and probabilistic preference, provides a more flexible technique to express DMs' relative preferences among states and is more comprehensive than existing models. A real-world water resource dispute is used to illustrate the process of identifying the hybrid preference structure. Discussions are made about the valuable strategic insights that are obtained when strength and probabilistic preferences are combined as a hybrid preference structure in conflict analysis.

Keyword: Graph model · Hybrid preference structure · Strength of preference · Probabilistic preference · Conflict Analysis

1 Introduction

Strategic conflicts – situations where multiple decision-makers (DMs) with differing perspectives interact – are common in the real world, particularly in the realm of environmental issues. The Graph Model for Conflict Resolution (GMCR) (Kilgour et al., 1987; Fang et al., 1993) is a widely used methodology in conflict analysis due to its simplicity and flexibility. Since DMs often have different preference structures, which influence conflict analysis and stability outcomes, various studies have refined GMCR to capture these dynamics. Kilgour et al. (1987) introduced crisp preferences, assuming all relative preferences are known. Later, Hamouda et al. (2004; 2006) added three-level preference strength, and Xu et al. (2009a) extended this to a multi-level preference structure.

Uncertainty in preference information can be modeled using various structures. Li et al. (2002; 2004) introduced unknown preferences for cases where there is no knowledge regarding DM's preference information. Grey numbers can also be used to model preference uncertainty (Kuang et al., 2013; 2015). Bashar et al. (2010; 2012) presented fuzzy preferences to model situations in which DMs might not have a clear-cut preference between two states. Bashar et al. (2016; 2017) further refined fuzzy preference with interval fuzzy preferences. Wu et al. (2020) introduced hesitant fuzzy preferences, while Yu and Deng (2024) proposed intuitionistic fuzzy preferences. Rêgo and Santos (2015) developed a probabilistic preference structure, later extending it to address vagueness in preference elicitation (Rêgo and Santos, 2018). When a single preference structure is insufficient, hybrid preferences integrate different types for a more nuanced conflict representation. Xu et al. (2008; 2009b) combined preference strength with unknown preference, while Yu et al. (2017; 2018) merged multi-level strength with fuzzy preferences for two or multiple DMs. Yu et al. (2019) further developed a framework incorporating both unknown and fuzzy preferences.

Although GMCR includes various preference structures, some conflicts are not sufficiently represented. For instance, in a water pollution dispute, an environmental agency may prefer that an industrial enterprise does not seriously pollute a nearby river into which it discharges waste. On the other hand, there may be different types of industrial enterprises, such as industries with no environmental concern and industries concerned about environmental impact issues. Furthermore, assume that each type of enterprise may have different preference levels regarding the states. For example, the environmentally concerned industry may greatly prefer not being fined by the environmental agency while the least concerned DM type may only mildly prefer such a situation. A hybrid preference structure could model these variations, integrating preference strength (Hamouda et al., 2004; 2006) and probabilistic preference (Rêgo and Santos, 2015; Rêgo and Vieira, 2021) for a more accurate representation.

Therefore, while previous research has examined strength of preference and probabilistic preference separately, a key limitation remains: DMs often exhibit both preference structures simultaneously. Existing models overlook this complexity, resulting in a limited representation of strategic conflicts. This paper addresses this gap by proposing a hybrid preference model that integrates both structures, offering a more accurate depiction of real-world conflicts and creating a more flexible graph model. With this in mind, this paper seeks to develop a novel hybrid preference model that combines preference strength and probabilistic preferences in conflicts involving two DMs, allowing them to express mild or strong preferences within a probabilistic framework.

The remainder of this paper is structured as follows: in Sect. 2, under the framework of GMCR with two DMs, a hybrid preference structure, which combines both three-level preference strength and probabilistic preference, is proposed. Furthermore, four graph model stability definitions (solution concepts) α-Nash, α, β-General Metarationality (GMR), α, β-Symmetric Metarationality (SMR), and α, β, γ- Sequential Stability (SEQ) are extended to the new hybrid preference structure. In Sect. 3, a real-world water dispute is considered to illustrate the process of identifying the hybrid preference structure in practice. The final section includes conclusions and directions for future research.

2 GMCR Under Strength of Preference and Probabilistic Preference for Two DMs

In a GMCR framework, it is normally assumed that preferences are known. If there is any degree of uncertainty, different existing types of preference structures, such as unknown, grey, fuzzy, or probabilistic preferences, may represent this uncertainty according to the type or level of information available.

Rêgo and Santos (2015) introduced an extension of the GMCR, called GMCRP, to model situations where DMs may have probabilistic preferences between states. In this framework, each DM is assumed to be able to specify the probability, $P(s,q)$, that the DM prefers state s over state q for every pair of states s and q within a conflict. The underlying concept is that individuals can use historical data to establish a probability distribution over the states. In certain conflict scenarios, DMs – who may include individuals, enterprises, or countries – might have multiple representatives, with each representative having a certain probability of being selected as the DM for a given conflict. While each representative may have a deterministic preference relation over the set of states, prior to selecting the actual representative, the DM's preference cannot be expressed using a standard preference relation. Instead, a probabilistic preference relation is used, where the probability that the DM prefers state s to state q is the sum of the probabilities assigned to the representatives who prefer s over q (Rêgo and Santos, 2015).

While probabilistic preferences can model the likelihood that a DM strictly prefers state s over q when choosing between them, they cannot account for situations involving three-level strengths of preference $\{\gg_i, >_i, \text{or} \sim_i\}$, where $i \in N$ denotes DM i and N is the set of DMs in a conflict. However, in real-world conflicts, some DMs' preferences over feasible states may be better represented by strong ($\gg_i$) or mild ($>_i$) relations (Hamouda et al., 2004; 2006), while others require probabilistic modeling. Therefore, a new hybrid preference structure within GMCR that integrates three-level preference strengths and probabilistic preferences to better capture these aspects and lead to a more realistic conflict representation is needed.

In this section, the probabilistic preference structures are extended to combine with three-levels of preference strength. Our objective is to propose an extension for the GMCRP (Rêgo and Santos, 2015) which also allows DMs to have strength of preference over the set of possible states. This hybrid preference structure may be used to model either a single DM who vacillates in a probabilistic fashion when choosing between states or a DM who is not monolithic.

2.1 Preference Structure Definition

By extending what Rêgo and Santos (2015) proposed as probabilistic preferences (GMCRP), in which DMs prefer one state over another with a certain probability, to develop the hybrid strength of preference and probabilistic preference relation, the notation $P_i(s,q)$ expresses the chance with which DM i mildly prefers state s to q. Such a probability is formally defined on $P_i : S \times S \to \mathbf{R}$, where S is the set of w feasible states in a conflict, and satisfies the following properties:

(a) $P_i(s, q) \geq 0, \forall s, q \in S$.
(b) $P_i(s, s) = 0, \forall s \in S$,
(c) $P_i(s > q) + P_i(q > s) \leq 1, \forall s, q \in S$.

In others words, (a) requires that for any two states in S, there is no negative probabilities. Condition (b) requires that no DM i can mildly prefer one state over itself with positive probability, and condition (c) requires that the sum of the probabilities that some DM i mildly prefers state s to q and mildly prefers q over s is at most equal to 1. This later condition allows DMs to be indifferent between states with positive probability. Hence, the difference $1 - P_i(s, q) - P_i(q, s)$ represents the probability with which DM i is indifferent between s and q.

Example 1: we present an adapted version of the hypothetical conflict used by Rêgo and Santos (2015) which illustrates the usability of GMCRP. Table 1 shows the DMs and their options, as well as feasible states of the conflict.

Table 1. Feasible states of the conflict in example 1.

DMs	Options	State number			
		s_1	s_2	s_3	s_4
Environmentalists (DM 1)	1. Be proactive	Y	N	Y	N
	2. Be reactive	N	Y	N	Y
Developers (DM 2)	3. Be sustainable	Y	Y	N	N
	4. Be unsustainable	N	N	Y	Y

"Y" means the respective option is selected and "N", otherwise

Let us assume that the preferences of DMs are deterministic, that DM 1's preferences are such that $(s_1 >_1 s_2 >_1 s_3 >_1 s_4)$ and that there are two types of DM 2, one who gives low priority to sustainable practices DM_{2L} and the other who has a high sense of sustainability DM_{2H}. The preferences of DM_{2L} and DM_{2H} are such that $(s_4 \gg_{2L} s_3 >_{2L} s_2 \gg_{2L} s_1)$ and $(s_2 >_{2H} s_1 >_{2H} s_4 \gg_{2H} s_3)$, respectively. We also consider that, if DM i deterministically mildly prefers state s to q, then we say that $P_i(s, q) = 1$. In this case, DM i mildly prefers s to q. Otherwise, we say that this probability is zero. Tables 2 and 3 show the probabilistic mild preferences of DM 1 and DM 2. Since DM 1's preferences are all certain, this DM's probabilities will be 1 or 0. For example, $P_1(s_1, s_2) = 1$, thus $P_1(s_2, s_1) = 0$.

In order to determine the probabilistic preferences for DM 2, as shown in Table 3, assume that there exists a probability distribution over the DM 2 types according to $P(DM2 = DM_{2L}) = 0.20$ and $P(DM2 = DM_{2H}) = 0.80$. Thus, the probability with which DM 2 mildly prefers s to q is given by the sum of the probabilities of the DM 2 types who at least mildly prefer s to q. For two cases in Table 3 there is a certain mild preference between states, such that $P_2(s_2, s_1) = 0.20 + 0.80 = 1$ and $P_2(s_4, s_3) = 0.20 + 0.80 = 1$, since both types of DM 2 at least mildly prefer s_2 to s_1 and s_4 to s_3, respectively.

Table 2. Probabilistic preferences for DM 1 from example 1.

DM 1	s_1	s_2	s_3	s_4
s_1	0.00	1.00	1.00	1.00
s_2	0.00	0.00	1.00	1.00
s_3	0.00	0.00	0.00	1.00
s_4	0.00	0.00	0.00	0.00

Table 3. Probabilistic preferences for DM 2 from example 1.

DM 2	s_1	s_2	s_3	s_4
s_1	0.00	0.00	0.80	0.80
s_2	1.00	0.00	0.80	0.80
s_3	0.20	0.20	0.00	0.00
s_4	0.20	0.20	1.00	0.00

The key assumption is that the probability distribution over DM types is known or can be estimated. If additional dimensions (e.g., DM types) are introduced, the method remains applicable as long as the probabilities for each type can be specified, since the final process of determining the preference between two states will always be given by the sum of the partial probabilities. For instance, the probability with which a specific DM mildly prefers state s over q is given by the sum of the probabilities of all existing DM types (i.e., all dimensions) that mildly prefer s to q.

In this paper, the probabilistic preference relation is extended to combine with three-levels of preference strength. In order to establish if any of the certain mild preferences obtained in Table 3 may actually be strong preferences, we have to define the new preference structure. The structure of the new hybrid preferences is defined as below:

Definition 1 (Strength of Preference and Probabilistic Preference): DM i's strength of preference and probabilistic preference information on S is a relation over S, denoted by a matrix $\mathcal{R}_i^S = (r_{s,q})_{w \times w}$, where the mapping function is $\mu_\mathcal{R} : S \times S \to P_i(s,q) \bigcup \{-2, -1, 0, 1, 2\}$, and $r_{s,q}$ represents the preference degree that state s is to be preferred to q, and $P_i(s,q)$ satisfies the conditions (a), (b) and (c).

The above preference structure includes both probabilistic preference structure proposed by Rêgo and Santos (2015) as well as the three-level strength of preference structure stablished by Hamouda et al. (2004; 2006). When there are no -2 and 2 in the conflict, the strength of preference and probabilistic preference structure is actually the probabilistic preference structure proposed by Rêgo and Santos (2015).

The interpretations of preference degree $r_{s,q}$ are as follows:

- $r_{s,q} = 2$ means that s is strongly preferred to q, i.e., $s \gg_i q$;

- $r_{s,q} = 1$ means that s is mildly preferred to q, i.e., $s>_i q$;
- $r_{s,q} \in (0,1) =$ means that s is likely mildly preferred to q, i.e., $P_i(s>_i q)$;
- $r_{s,q} = 0$ means that s is equally preferred to q, i.e., $s\sim_i q$;
- $r_{s,q} = -1$ means that q is mildly preferred to s, i.e., $q>_i s$.
- $r_{s,q} = -2$ means that q is strongly preferred to s, i.e., $q\gg_i s$.

Taking $S = \{s_1, s_2, s_3, s_4\}$ from *example 1*, the matrix $\mathcal{R}^S_{DM2}$ in Table 4 describes DM 2's strength of preference and probabilistic preference combination, as proposed in this paper. For instance, the number "2" in the fourth row and third column of $\mathcal{R}^S_{DM2}$ means that the preference of state s_4 over s_3 for DM 2 is 2, which indicates that s_4 is strongly preferred to s_3, i.e., $s_4 \gg_2 s_3$. To identify this preference relation, one needs to verify all certain probabilities from Table 3, i.e., $P_i(s, q) = 1$, then, it is necessary to identify in which partial probabilities both represent strong preference between two states. For instance, only $P_2(s_2, s_1) = 1$ and $P_2(s_4, s_3) = 1$ in Table 3. For the first case, the partial probabilities show that $(s_2 \gg_{2L} s_1)$ for DM_{2L}, but $(s_2 >_{2H} s_1)$ for DM_{2H}. Thus, even if there is a mild preference certainty between s_2 and s_1, there is no strong preference certainty for both types of DM 2. On the other hand, for the second case $P_2(s_4, s_3) = 1$, the partial probabilities represent both strong preference for $DM_{2L}(s_4 \gg_{2L} s_3)$, and for DM_{2H} $(s_4 \gg_{2H} s_3)$, thus DM 2 certainly strongly prefers s_4 to s_3, i.e., $(s_4 \gg_{DM2} s_3)$.

Table 4. DM 2's hybrid preference combination from example 1.

$\mathcal{R}^S_{DM2} =$		s_1	s_2	s_3	s_4
	s_1	0	0	0.8	0.8
	s_2	1	0	0.8	0.8
	s_3	0.2	0.2	0	-1
	s_4	0.2	0.2	2	0

2.2 Stability Definitions for Two DM Conflicts

In this subsection, we develop new solution concepts for GMCR considering strength and probabilistic preferences. Consider parameters α, β, and γ lying in the interval $[0, 1)$. We define a γ-improvement for DM i from a given state s to be the set $R_i^{+\gamma}(s) = \{q \in R_i(s) : r_{q,s} > \gamma\}$. We denote by $\phi_i^{+\gamma}(s)$ the set of all states that i prefers to s with preference degree greater than γ, i.e., $\phi_i^{+\gamma}(s) = \{q \in S : r_{q,s} > \gamma\}$. Note that $R_i^{+\gamma}(s) = \phi_i^{+\gamma}(s) \cap R_i(s)$. Obviously, for all cases in which there is certainty of preference between states, i.e., $r_{q,s} = 1$ or 2, the value of γ is irrelevant, however, γ plays an important role when it comes to preferences that involve probability, thus γ is interpreted as the threshold for the preference of the DM with strength and probabilistic preferences.

2.2.1 General Stability Definitions

The concepts of general, strong and weak stability definitions proposed by Hamouda et al., (2004, 2006) are considered to develop the stability definitions for the hybrid strength of preference and probabilistic preference relations. Firstly, general stabilities are defined, and then the two subclasses, strong and weak, are determined. General stabilities in this hybrid context, are not the same as the stability definitions of probabilistic preference (Rêgo and Santos, 2015; Rêgo and Vieira, 2021) because in their case they do not take into account the strength of preference. Furthermore, strong and weak stability definitions are developed to reflect the additional strength of preference information contained in the preference degree.

Definition 2: A state $s \in S$ is α-Nash stable (α-R) for DM $i \in N$ iff $R_i^{+(1-\alpha)}(s) = \varnothing$.

Then a state s is α-Nash stable for DM i if among all states that i can achieve from s, there is **no** state that i prefers to s with preference degree greater than $1 - \alpha$.

Definition 3: A state $s \in S$ is α, β-General GMR ((α, β)-GGMR) stable for DM $i \in N$ iff for every $s_1 \in R_i^{+(1-\alpha)}(s)$, there exists $s_2 \in R_j(s_1)$ such that $r_{s2,s}^i \leq 1 - \beta$.

Definition 4: A state $s \in S$ is α, β-General SMR ((α, β)-GSMR) stable for DM $i \in N$ iff for every $s_1 \in R_i^{+(1-\alpha)}(s)$, there exists $s_2 \in R_j(s_1)$ such that $r_{s2,s}^i \leq 1 - \beta$ and. $r_{s3,s}^i \leq 1 - \alpha$, for every $s_3 \in R_i(s_2)$.

Definition 5: A state $s \in S$ is α, β, γ-General SEQ ((α, β, γ)-GSEQ) stable for DM $i \in N$ iff for every $s_1 \in R_i^{+(1-\alpha)}(s)$, there exists $s_2 \in R_j^{+\gamma}(s_1)$ such that $r_{s2,s}^i \leq 1 - \beta$.

Similar to the solution concepts presented by Rêgo and Santos (2015) and Rêgo and Vieira (2021), intuitively, if a state s is (α, β-GGMR) stable for DM i, he or she has no incentive to move away from it, because for all state s_1 that i at least mildly prefers over s with preference degree greater than $1 - \alpha$, there exists a reachable state s_2 for the opponent of i, such that i does not mildly prefer s_2 over s with preference degree greater than $1 - \beta$. Besides that, in an (α, β-GSMR) stable state for DM i, DM i cannot escape from this latter situation for a state that he or she at least mildly prefers over s with preference degree greater $1 - \alpha$. In an (α, β, γ-GSEQ) stable state for DM i, all the moves in the reaction of the opponent of DM i are unilateral γ-improvements, but no requirement to whether DM i may counter-react is made.

2.2.2 Strong and Weak Stability Definitions

When strength of preference is introduced into the graph model, stability definitions can be strong or weak, according to the degree of sanctioning. As discussed in Hamouda et al. (2004, 2006) strong and weak stabilities only include GMR, SMR, and SEQ because Nash does not involve sanctions, the same lies for a probabilistic context.

Definition 6: A state $s \in S$ is α, β-Strongly GMR ((α, β)-SGMR) stable for DM $i \in N$ iff for every $s_1 \in R_i^{+(1-\alpha)}(s)$, there exists $s_2 \in R_j(s_1)$ such that $r_{s2,s}^i = -2$.

Definition 7: A state $s \in S$ is α, β-Strongly SMR ((α, β)-SSMR) stable for DM $i \in N$ iff for every $s_1 \in R_i^{+(1-\alpha)}(s)$, there exists $s_2 \in R_j(s_1)$ such that $r_{s2,s}^i = -2$ and $r_{s3,s}^i = -2$, for every $s_3 \in R_i(s_2)$.

Definition 8: A state $s \in S$ is α, β, γ-Strongly SEQ ((α, β, γ)-SSEQ) stable for DM $i \in N$ iff for every $s_1 \in R_i^{+(1-\alpha)}(s)$, there exists $s_2 \in R_j^{+\gamma}(s_1)$ such that $r_{s2,s}^i = -2$.

Definition 9: Let $s \in S$ and $i \in N$. A state s is weakly stable for DM i iff s is general stable, but not strongly stable for some stability definition.

In this paper, (α, β, γ-SSEQ) is different from what is discussed in Rêgo and Vieira (2016). In their work, it is a definition of Symmetric Sequential Stability (SSEQ). In this paper, as discussed previously, (α, β, γ-SSEQ) refers to strongly sequential stability.

3 Real-World Case Application in a Two-DM Conflict

In the textile industry, washing in finishing operations enhances product quality but can cause pollution. Inefficient dyeing and washing processes release up to 200,000 tons of dyes into effluents annually worldwide (Chequer et al., 2013).

In Brazil, these finishing operations of washing are usually performed by textile laundries, which generate many tailings that can cause environmental problems in the regions where such companies are located. In this context, the Agreste region of Pernambuco State, which is the second largest textile manufacturer in the country, has suffered from the environmental impacts caused by the activities carried out in the textile laundries causing river coloration and fish kill due to polluting effluents. However, environmental legislation obliges these textile industries to operate wastewater treatment from their dye-containing effluents, before disposal into water bodies (CPRH, 2025). Figure 1 shows the environmental impact of textile laundries in one of the most important rivers in Pernambuco State, where the majority of the 800 existing laundries dump blue dye from the jeans industry into the Capibaribe River.

Fig. 1. Chemical effluents being dumped into the Capibaribe River.

The Capibaribe River plays an extremely important historical, social, and economic role in the formation and progress of Pernambuco State and its capital, Recife. It is the largest river in Pernambuco State. It runs for two hundred and fifty kilometers, passing through forty-two cities. Along its route, the Capibaribe River receives contributions from approximately 72 tributaries, in a basin with an area of 5,880 square kilometers (Fig. 2). Inspection and control operations are performed by the State Agency for the Environment and Water Resources of Pernambuco State which has the power to fine irregular laundries or even shut down illegal ones (Silva et al., 2022).

Fig. 2. Capibaribe River location in Pernambuco State, Brazil.

3.1 Conflict Modeling

In this subsection we apply the hybrid preference structure proposed in this paper in the water pollution dispute previously discussed.

3.1.1 DMs, Options and Feasible States

Let us consider the following real-world environmental conflict. In a water pollution dispute, an Environmental Agency (DM 1) has the role of inspecting Laundry Industries (DM 2) that operate chemical processes in the textile sector. DM 1 may fine (option 1) or shut down (option 2) laundries that are not environmentally disposing of its effluents from their industrial processes. On the other hand, the Textile Laundry (DM 2) may adopt a sustainable behavior and treat its effluents (option 3), or not treat them (not option 3), thus causing a major environmental impact on water bodies. Since there are three options ($2^3 = 8$), there is a total of 8 possible states and 6 feasible states (Option 2 cannot be taken if option 3 is selected). Table 5 presents the feasible combinations of states in this conflict. These states are the vertices (nodes) shown in Fig. 3, which represents the graph model for the water pollution dispute described above.

3.1.2 Preferences

Since the options available for DM 1 are fine and shut down, let us consider that DM 1 is in s_1 (Fig. 3), for instance, this DM can choose to remain in s_1 (Y-Y-N) or move to s_3 (Y-N-N) if he/she chooses not to shut down the laundry (not option 2). DM 1 can also make a state transition from s_1 to s_4 (N-Y-N), if he/she decides, in that case, not to

Strength of Preference and Probabilistic Preference in the Graph Model

Table 5. Feasible states of the water pollution dispute in option form.

DMs	Options	State number					
		s_1	s_2	s_3	s_4	s_5	s_6
Environmental agency (DM 1)	1. Fine	Y	Y	Y	N	N	N
	2. Shut down	Y	N	N	Y	N	N
Textile laundry (DM 2)	3. Treat	N	Y	N	N	Y	N

fine (not option 1) the laundry. Thus, the set $R_{DM1}(s)$, considering $s = (s_1)$, is made of states s_3 and s_4. Let us consider now that DM 2, e.g., is in state s_3 (Y-N-N). This DM can remain in that state or move to s_2 (Y-N-Y) if he/she chooses to treat the effluents (option 3). Then, $R_{DM2}(s)$, considering $s = (s_3)$, is made of only state s_2. The other achievable sets of unilateral moves are obtained in a similar manner (Fig. 3). The conflict is modeled as having an initial state and its development depends on the DMs changing actions and therefore states.

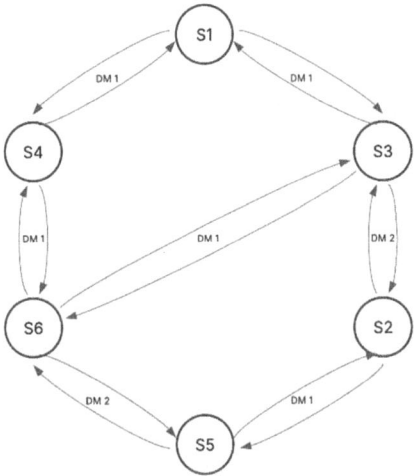

Fig. 3. Graph model for the water dispute showing all DMs' possible moves.

For this conflict, we assume that DM 1 has deterministic preferences and that there are two types of DM 2, one who is concerned about environmental impact issues (DM_{2C}) and the other type who is environmentally unconcerned (DM_{2U}). Preference rankings for each DM are presented as follows:

- $DM_1: s_5 >_1 s_2 >_1 s_1 >_1 s_4 >_1 s_3 >_1 s_6$
- $DM_{2C}: s_5 >_{2C} s_6 \gg_{2C} s_2 >_{2C} s_3 \gg_{2C} s_4 >_{2C} s_1$
- $DM_{2U}: s_6 >_{2U} s_3 >_{2U} s_5 >_{2U} s_2 \gg_{2U} s_4 >_{2U} s_1$

We also consider that, if DM i deterministically mildly prefers state s to q, then we say that $P_i(s, q) = 1$. In this case, DM i mildly prefers s to q. Otherwise, we say that this probability is zero. Tables 6 and 7 show the probabilistic mild preferences of DM 1 and DM 2. Since DM 1's preferences are all certain, this DM's probabilities will be 1 or 0. For instance, $P_1(s_5, s_2) = 1$, thus $P_1(s_2, s_5) = 0$.

Table 6. Probabilistic preferences of the environmental agency (DM 1).

DM 1	s_1	s_2	s_3	s_4	s_5	s_6
s_1	0.00	0.00	1.00	1.00	0.00	1.00
s_2	1.00	0.00	1.00	1.00	0.00	1.00
s_3	0.00	0.00	0.00	0.00	0.00	1.00
s_4	0.00	0.00	1.00	0.00	0.00	1.00
s_5	1.00	1.00	1.00	1.00	0.00	1.00
s_6	0.00	0.00	0.00	0.00	0.00	0.00

Table 7. Probabilistic preferences of the textile industry (DM 2).

DM 2	s_1	s_2	s_3	s_4	s_5	s_6
s_1	0.00	0.00	0.00	0.00	0.00	0.00
s_2	1.00	0.00	0.40	1.00	0.00	0.00
s_3	1.00	0.60	0.00	1.00	0.60	0.00
s_4	1.00	0.00	0.00	0.00	0.00	0.00
s_5	1.00	1.00	0.40	1.00	0.00	0.40
s_6	1.00	1.00	1.00	1.00	0.60	0.00

In order to determine the probabilistic preferences for DM 2, as shown in Table 7, assume that there exists a probability distribution over the DM 2 types according to $P(DM\,2 = DM_{2C}) = 0.40$ and $P(DM\,2 = DM_{2U}) = 0.60$. Thus, the probability with which DM 2 mildly prefers s to q is given by the sum of the probabilities of the DM 2 types who at least mildly prefer s to q. For some cases in Table 7 there is a certain mild preference between states, such as $P_2(s_2, s_1) = 0.40 + 0.60 = 1$ and $P_2(s_3, s_4) = 0.40 + 0.60 = 1$, since both types of DM 2 at least mildly prefer s_2 to s_1 and s_3 to s_4, respectively.

The matrix $\mathcal{R}_{DM2}^S$ in Table 8 describes DM 2's strength of preference and probabilistic preference combination, as proposed in this paper. For instance, the number "2" in the fifth row and first column of $\mathcal{R}_{DM2}^S$ means that the preference of state s_5 over s_1 for DM 2 is 2, which indicates that s_5 is strongly preferred to s_1, i.e., $s_5 \gg_2 s_1$. To identify this preference relation, one needs to verify all certain probabilities from Table 7, i.e., $P_i(s, q) = 1$, then, it is necessary to identify in which partial probabilities from DM 2 types' preference rankings both represent strong preference between two states. For instance, let us consider $P_2(s_6, s_2) = 1$ and $P_2(s_5, s_4) = 1$ in Table 7. For the first case, the partial probabilities show that $(s_6 \gg_{2C} s_2)$ for DM_{2C}, but $(s_6 >_{2U} s_2)$ for DM_{2U}. Thus, even if there is a mild preference certainty between s_6 and s_2, there is no strong preference certainty for both types of DM 2, then the respective value for the preference relation between s_6 and s_2 remains as "1" in Table 8. On the other hand, for the second case $P_2(s_5, s_4) = 1$, the partial probabilities represent both strong preference for $DM_{2C}(s_5 \gg_{2C} s_4)$, and for DM_{2U} $(s_5 \gg_{2U} s_4)$, thus DM 2 certainly strongly prefers s_5 to s_4, i.e., $(s_5 \gg_{DM2} s_4)$, then the respective value from this preference relation is "2" in Table 8. For the others certain values, i.e., $P_i(s, q) = 1$ in Table 7 similar evaluation has to be done.

Table 8. DM 2's hybrid preference combination.

$\mathcal{R}^S_{DM2} =$		s_1	s_2	s_3	s_4	s_5	s_6
	s_1	0	−2	−2	−1	−2	−2
	s_2	2	0	0.4	2	−1	−1
	s_3	2	0.6	0	2	0.6	−1
	s_4	1	−2	−2	0	−2	−2
	s_5	2	1	0.4	2	0	0.4
	s_6	2	1	1	2	0.6	0

4 Conclusions

In this paper, we extended ideas used in Rêgo and Santos (2015) to propose a hybrid preference structure that combines both strength of preference and probabilistic preference structures within GMCR. Four stability definitions (solution concepts) α-Nash, α, β-General Metarationality, α, β-Symmetric Metarationality, and α, β, γ-Sequential Stability were extended to include general, strong and weak stabilities for conflicts with two DMs. The proposition presented in this paper combines the flexibility in integrating different preference features of DMs for real-world conflicts through strength and probabilistic preference structures and the representations of solution concepts for obtaining equilibria in a more realistic manner.

Calculating equilibria within GMCR is crucial because it helps identify the stable strategic choices for DMs in a conflict. The equilibria provide insights into how the conflict might evolve and what potential resolutions are feasible based on the preferences and strategies of the involved parties. In this paper, due to paper length limitations, we could not provide stable states for the case study. For further research the equilibria result for the real-world case application is necessary for a more comprehensive illustration of the proposed hybrid preference structure.

It would also be interesting to compare the robustness of equilibria by comparing the parameter ranges for which a particular state satisfies a given stability definition.

This combined preference structure can be extended for conflicts with multiple DMs. Moreover, for conflicts with many states, calculation complexity may increase, but this is a practical challenge rather than a theoretical limitation of the framework. For this, a matrix representation can be developed to facilitate stability analysis, making the model computationally efficient and scalable for complex conflicts.

This approach differs from existing hybrid preference structures by incorporating probabilistic contexts rather than relying on fuzzy or unknown preferences explored in previous studies. Probabilistic preference models the likelihood of DMs preferring one state over another, a feature not explicitly captured in fuzzy-based models. Additionally, the proposed model introduces a three-level strength of preference, allowing DMs to express strong or mild preferences, leading to a more nuanced and realistic conflict representation within GMCR. This hybrid structure better reflects real-world decision-making, where DMs may exhibit both deterministic preference strengths and probabilistic uncertainty in strategic conflicts.

Acknowledgements. The authors would like to acknowledge the financial support of the Fundação de Amparo à Ciência e Tecnologia de Pernambuco (FACEPE) by grant IBPG-1505-3.08/20, the partial financial support of the Coordenação de Aperfeiçoamento de Pessoal de Nível Superior (CAPES) – Finance code 001, and the general support of the Conselho Nacional de Desenvolvimento Científico e Tecnólogico (CNPq).

References

Bashar, M.A., Hipel, K.W., Kilgour, D.M.: Fuzzy preferences in a two-decision maker graph model. In: IEEE International Conference on Systems, Man and Cybernetics (2010). https://doi.org/10.1109/icsmc.2010.5641995

Bashar, M.A., Kilgour, D.M., Hipel, K.W.: Fuzzy preferences in the graph model for conflict resolution. IEEE Trans. Fuzzy Syst. **20**(4), 760–770 (2012). https://doi.org/10.1109/tfuzz.2012.2183603

Bashar, M.A., Obeidi, A., Kilgour, D.M., Hipel, K.W.: Modeling fuzzy and interval fuzzy preferences within a graph model framework. IEEE Trans. Fuzzy Syst. **24**(4), 765–778 (2016). https://doi.org/10.1109/tfuzz.2015.2446536

Bashar, M.A., Hipel, K.W., Kilgour, D.M., Obeidi, A.: Interval fuzzy preferences in the graph model for conflict resolution. Fuzzy Optim. Decis. Making **17**(3), 287–315 (2017). https://doi.org/10.1007/s10700-017-9279-7

Chequer, F.M.D., de Oliveira, G.A.R., Anastacio Ferraz, E.R., Carvalho, J., Boldrin Zanoni, M.V., de Oliveir, D.P.: Textile dyes: dyeing process and environmental impact. Eco-Friendly Textile Dyeing and Finishing (2013). https://doi.org/10.5772/53659

CPRH (State Agency for the Environment and Water Resources) (2025). http://www.cprh.pe.gov.br. Accessed 10 Jan 2025

Fang, L., Hipel, K.W., Kilgour, D.M.: Interactive Decision Making: The Graph Model for Conflict Resolution. Wiley, New York (1993)

Hamouda, L., Kilgour, D.M., Hipel, K.W.: Strength of preference in the graph model for conflict resolution. Group Decis. Negot. **13**(5), 449–462 (2004). https://doi.org/10.1023/b:grup.0000045751.2120

Hamouda, L., Kilgour, D.M., Hipel, K.W.: Strength of preference in graph models for multiple-decision-maker conflicts. Appl. Math. Comput. **179**(1), 314–327 (2006). https://doi.org/10.1016/j.amc.2005.11.109

Kilgour, D.M., Hipel, K.W., Fang, L.: The graph model for conflicts. Automatica **23**(1), 41–55 (1987)

Kuang, H., Hipel, K.W., Kilgour, D.M., Bashar, M.A.: A case study of grey-based preference in a graph model for conflict resolution with two decision makers. In: 2013 IEEE International Conference on Systems, Man, and Cybernetics (2013). https://doi.org/10.1109/smc.2013.349

Kuang, H., Bashar, M.A., Hipel, K.W., Kilgour, D.M.: Grey-based preference in a graph model for conflict resolution with multiple decision makers. IEEE Trans. Syst. Man Cybern. Syst. **45**(9), 1254–1267 (2015). https://doi.org/10.1109/tsmc.2014.2387096

Li, K.W., Hipel, K.W., Kilgour, D.M., Fang, L.: Stability definitions for 2-player conflict models with uncertain preferences. In: IEEE International Conference on Systems, Man and Cybernetics (2002). https://doi.org/10.1109/icsmc.2002.1175659

Li, K.W., Hipel, K.W., Kilgour, D.M., Fang, L.: Preference uncertainty in the graph model for conflict resolution. IEEE Trans. Syst. Man Cybern. Part A Syst. Hum. Part A Syst. Hum. **34**(4), 507–520 (2004). https://doi.org/10.1109/tsmca.2004.826282

Rêgo, L.C., dos Santos, A.M.: Probabilistic preferences in the graph model for conflict resolution. IEEE Trans. Syst. Man Cybern. Syst. **45**(4), 595–608 (2015). https://doi.org/10.1109/tsmc.2014.2379626

Rêgo, L.C., dos Santos, A.M.: Upper and lower probabilistic preferences in the graph model for conflict resolution. Int. J. Approx. Reason. **98**, 96–111 (2018). https://doi.org/10.1016/j.ijar.2018.04.008

Rêgo, L.C., Vieira, G.I.A.: Symmetric sequential stability in the graph model for conflict resolution with multiple decision makers. Group Decis. Negot. **26**(4), 775–792 (2016). https://doi.org/10.1007/s10726-016-9520-8

Rêgo, L.C., Vieira, G.I.A.: Matrix representation of solution concepts in the graph model for conflict resolution with probabilistic preferences and multiple decision makers. Group Decis. Negot. **30**(3), 697–717 (2021). https://doi.org/10.1007/s10726-021-09729-y

Silva, B.L., Xavier, M.G.P., Sobral, M.F.F.: Jeans processing and water reuse systems in a brazilian semiarid region. Revista De Gestão Social E Ambiental **16**(2), e03027 (2022). https://doi.org/10.24857/rgsa.v16n2-030

Wu, N., Xu, Y., Kilgour, D.M., Fang, L.: Composite decision makers in the graph model for conflict resolution: hesitant fuzzy preference modeling. IEEE Trans. Syst. Man Cybern. Syst. 1–14 (2020). https://doi.org/10.1109/tsmc.2020.2992272

Xu, H., Hipel, K.W., Kilgour, D.M.: Preference strength and uncertainty in the graph model for conflict resolution for two decision-makers. In: 2008 IEEE International Conference on Systems, Man and Cybernetics (2008). https://doi.org/10.1109/icsmc.2008.4811739

Xu, H., Hipel, K.W., Kilgour, D.M.: Multiple levels of preference in interactive strategic decisions. Discret. Appl. Math. **157**(15), 3300–3313 (2009a). https://doi.org/10.1016/j.dam.2009.06.032

Xu, H., Hipel, K.W., Kilgour, D.M., Chen, Y.: Combining strength and uncertainty for preferences in the graph model for conflict resolution with multiple decision makers. Theor. Decis. **69**(4), 497–521 (2009b). https://doi.org/10.1007/s11238-009-9134-6

Yu, J., Hipel, K.W., Kilgour, D.M., Fang, L.: Fuzzy strength of preference in the graph model for conflict resolution with two decision makers. In: 2017 IEEE International Conference on Systems, Man, and Cybernetics (SMC) (2017). https://doi.org/10.1109/smc.2017.8123186

Yu, J., Hipel, K.W., Kilgour, D.M., Fang, L.: Fuzzy levels of preference strength in a graph model with multiple decision makers. Fuzzy Sets Syst. (2018). https://doi.org/10.1016/j.fss.2018.12.016

Yu, J., Hipel, K.W., Kilgour, M., Fang, L., Yin, K.: Graph model under unknown and fuzzy preferences. IEEE Trans. Fuzzy Syst. 1 (2019). https://doi.org/10.1109/tfuzz.2019.2905222

Yu, J., Deng, X.: The graph model under intuitionistic fuzzy preference considering consensus and attitudes. IEEE Trans. Fuzzy Syst. Fuzzy Syst. **32**(7), 3914–3927 (2024). https://doi.org/10.1109/TFUZZ.2024.3385769

Balancing Affordability and Profitability in Urban Public Transport: Implications from Singapore and Hong Kong

Kai Xu[1](✉), Yu Maemura[2], and Kazumasa Ozawa[3]

[1] PADECO Co., Ltd., 6-17-19 Shinbashi, Minato-Ku, Tokyo, Japan
xu-kai@outlook.com
[2] The University of Tokyo, 5-1-5 Kashiwanoha, Kashiwa, Chiba, Japan
[3] National Graduate Institute for Policy Studies, 7-22-1 Roppongi, Minato-Ku, Tokyo, Japan

Abstract. Public transport can be considered both a public good and a private operation. This study focuses on the pricing dilemma that emerges from the need to balance affordability for the public and profitability for operators. Through a systematic review and ex-post evaluation of the pricing policies in Singapore and Hong Kong, we analyze the impacts of various transport policy schemes on the pricing dilemma to draw implications for transport policy making. Pricing policies are evaluated by integrating the theory of change and realist evaluations to observe how fare schemes, subsidy schemes, and finance schemes impact affordability and profitability. The case studies reveal that fare schemes may be less effective in addressing affordability compared with subsidy schemes targeting low-income groups. Finance schemes, on the other hand, set the financial structure of the public transport system and significantly influence the balance of the pricing dilemma and can be reformed to adapt to changing socioeconomic conditions.

Keywords: Public Transport · Ex-post Evaluation · Affordability · Profitability

1 Introduction

1.1 Pricing Policies and the Pricing Dilemma

The role of pricing in transport management has been a key topic for transportation managers and policy-makers since the birth of transport economics in the 1960s. Price can affect demand elasticity, and pricing at the equilibrium point can maximize economic gains. However, the public transport sector is clearly significantly more complicated than such theoretical assumptions, and the balancing of affordability for the user and profitability for the operator is a fundamental problem that public transport operators and policy makers must address.

Figure 1 illustrates the flow of capital among public transportation stakeholders. Fares are paid by individual users for the public transport service contributing to fare box revenue. The revenue can work as a financing source for the improvement of the public or privately operated transport services, and operators can be authorized to collect profits from the revenue or subsided if revenues are not sufficient to cover operation costs.

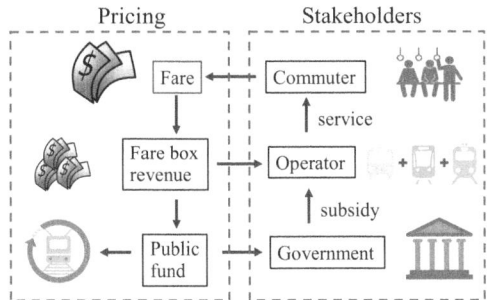

Fig. 1. Public transport pricing and stakeholders involved.

Regardless of the various institutional arrangements for transport services, "public" transport can only function if it is affordable for the majority of users. Public transport affordability is commonly defined and measured as the proportion of household expenditure utilized for public transport. While lower fares can ensure inclusion, fare revenue is also essential for maintaining the financial viability and quality of public transport operations. Fares must therefore be sufficiently high enough to guarantee profitability, especially when services are operated by a profit-driven private operator.

Urban public transport systems, to a large extent, are shaped by the interactions between pricing policy and cities that evolve and change the nature of the pricing dilemma. In this study, a series of pricing cases are reviewed to showcase modern practices, and to enhance the understanding on relationships and impacts of pricing policies on the pricing dilemma.

1.2 Review of Pricing Schemes

In this section, we describe various urban transport systems that have utilized three pricing schemes - fare schemes, subsidy schemes, and finance schemes - to balance affordability for commuters and financial sustainability for operators.

Fare Schemes. Fare schemes directly regulate fare levels to enhance affordability. In general, this includes setting the fare structure and the associated fare adjustments. Typically, three types of fare structures are present in the public transport field: flat rate, zone-based, or distance-based fare structures [1]. Fare adjustments allow operators to adjust the fare level by considering supply, demand, and capacity constraints. For example, the Manila Metro Rail Transit System Line 3 operates under a distance-based fare structure with fares ranging from 13–28 Philippine Pesos (equivalent to 0.25–0.5 US Dollars). The low fare is maintained by government subsidies with the primary goal to boost ridership. However, the low fares do not appear to generate sufficient revenue to cover operational and maintenance costs, resulting in heavy government subsidies and deteriorating service quality [2].

In contrast, cities like Luxembourg introduced free nationwide public transport in 2020 to shift commuters from private cars to public transport and reduce road congestion. This scheme improves the affordability but also places a financial burden on the government.

Subsidy Schemes. Subsidy schemes provide financial support from the government to the producer (builder or operator) and consumer (user) of the public transport sector [3]. There are demand-side and supply-side subsidies, with funding sources differing from taxes, cross-subsidies, or a combination of the two. For example, Brazil introduced a demand-side subsidy through a compulsory regulation on public transport vouchers in 1987. Employers purchased and distributed vouchers to employees who pay a 6% salary deduction to participate. The scheme caps commuting expenses at 6% of salary but is criticized for its misuse, as vouchers are often sold in secondary markets.

Similarly in Tokyo, most employers provide commuter allowances to cover public transport expenses. Employees are indirectly incentivized to live farther from city centers, impacting housing prices and urban expansion. Although such subsidies alleviate commuting costs for employees, it also indirectly affects wage levels and land use patterns.

Finance Schemes. Finance schemes regulate the role of public and private stakeholders and define who receives the revenue and pays the cost. The Manila Metro Rail Transit System Line 3 operates under a public-oriented finance scheme where the government sets up public funds to cover operational deficits. While this approach ensures affordability it burdens public funds. By contrast, Tokyo's private-oriented finance scheme allows private operators to set fares and be responsible for financial sustainability. This results in higher fares but ensures operators' profitability without government intervention.

1.3 Integrating the Schemes as a Policy Package

Figure 2 presents a diagram of the relationship between pricing schemes, the pricing dilemma, and public transport systems, with affordability and profitability highlighted as central elements requiring balance. While existing literature has demonstrated independent patterns in how a specific scheme interacted with the system, we aim to describe the complex dynamics of how the pricing policy as a package impacts the pricing dilemma.

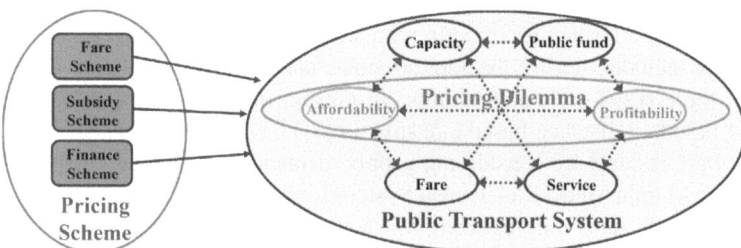

Fig. 2. Relationship between the pricing scheme, pricing dilemma, and public transport system.

The objective of this study is to investigate existing pricing practices in a systematic manner, from the outcomes of individual interventions to the impact policy packages on the pricing dilemma, and to extract evidence-based learnings for informed decisions regarding pricing policy. For this purpose, an analytical approach that combines the theory of change and the realist evaluation method was implemented, which also contributes to evaluation methodology. The findings produce practical implications for decision makers in the transportation sector, by providing general and systematic structure to the dynamic interactions between complex pricing policies and the public transportation system, that go beyond the consideration of the positive or negative impacts of individual pricing schemes on affordability and profitability.

2 Methods and Data: An Integrated Ex-post Evaluation of Transport Pricing Policy Packages

Previous studies proposed a variety of methods for transport policy evaluation, including multicriteria decision-making, indicator-based models, optimization, and simulations [4]. Significant potential and gaps remain for improving public transport pricing policy through systematic ex-post evaluations [5, 6]. In the face of such challenges, the theory of change and the realist evaluation method, which are classic evaluation methods in medical research [7] that focus on post-impact and associated causality analysis, have recently shown increasing applications in transport policy evaluation [8].

The theory of change is a systematic and cumulative study of the links between activities, outcomes, and the context of an intervention. The evaluation often adopts the form of an intervention logic model, which is a structured map to include key components and to demonstrate how the links between factors at various levels are combined to produce observed outcomes [9]. Realist evaluations combine quantitative and qualitative methods to investigate which combination of the mechanism and context is responsible for producing pragmatic and observable outcomes of an intervention and aim to develop and refine "context–mechanism–outcome" (CMO) configurations [10].

Blamey and Mackenzie compared the two methods and identified their roles in ex-post evaluations as "implementation theory" and "program theory", where the former provides review about policy implementation and the latter focuses on embedded causality [11]. The theory of change has a broader scope to ensure that the complex effects brought about by individual interventions can be recognized, whereas realist evaluations are focused on the entire policy package and mechanisms that produce observable outcomes.

Figure 3 illustrates the integration of the two approaches for ex-post evaluation of transport pricing policy in this study. Interventions are defined as a regulation or enforcement action. Outputs are the immediate effect of policy intervention, whereas the outcome and impact are the medium- and long-term effects. This information is integrated into a realist evaluation by considering how each individual scheme is linked to the context, mechanism, and outcome of the policy package. Context is defined as the features of the conditions in which the intervention is introduced and implemented and functions. Mechanism refers to the logical description of how an intervention creates a certain outcome under the context, thus connecting the context and intervention within the framework. Implications can be derived by testing and refining the revealed CMO sets so that they eventually become independent of specific cases and generally applicable.

Following the framework, key steps of ex-post evaluation are summarized in Table 1.

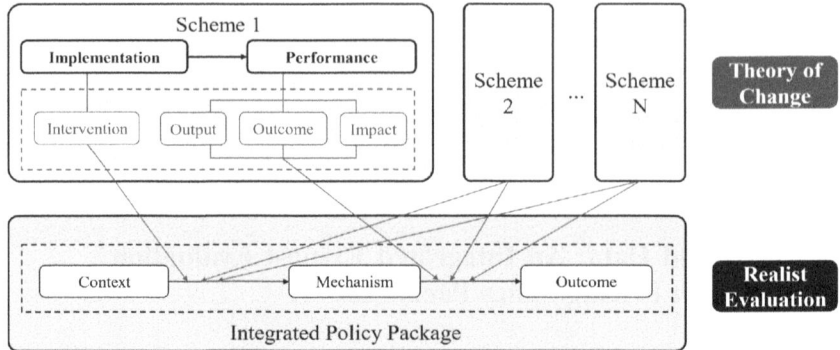

Fig. 3. Methodological framework.

Table 1. Steps involved in ex-post evaluation.

Method	Key step
Theory of change	1. Data collection on policy implementation and performance 2. Theory of change analysis on individual pricing scheme using the intervention logic model
Realist evaluation	3. Integration of intervention logic models results and summary of CMOs 4. Illustration by causality maps of the ex-post evaluation

Secondary data is collected and analyzed following the indicator sets suggested by Litman [12], Jeon [13], and Litman and Burwell [14]. Overall, the evaluation framework can deconstruct the compound effect of policies on the pricing dilemma based on quantitative indicators and qualitative evidence to abduct the underlying policy functioning mechanisms.

We select Singapore and Hong Kong as the cases for study. They are both home to modern megacities with relatively comparable populations, Gross Domestic Product levels, and even geographical conditions, such as limited urban areas and high traffic volumes. Their urban transport systems are highly recognized, with a modal share of public transport exceeding 80%. Both face tensions between affordability and profitability, and their divergent approaches provide a natural experiment to analyze how policy choices shape outcomes under different socioeconomic conditions. Both cities also provide a substantial amount of high-quality public documentation and data which enable us to achieve a level of saturation for each case. Detailed secondary data collected is listed in Appendix.

3 Ex-post Evaluation of the Pricing Policy in Singapore and Hong Kong

The study analyzes 10 pricing schemes in total from Singapore and Hong Kong.

3.1 Evaluating the Pricing Policy of Singapore

The initial theory of change analysis was performed to identify the context, outputs, and outcomes of the following pricing schemes (Table 2).

Table 2. Public transport pricing policy of Singapore.

Type	Pricing scheme	Brief description
Fare scheme	1. Distance-based fare scheme	Fare is charged based on the distance traveled regardless of public transport modes
	2. New capacity factor fare adjustment scheme	New capacity factor is included in the fare adjustment formula to reflect the network capacity cost
Subsidy scheme	3. Workfare transport concession scheme	To provide a lower concession fare (15% discount) for eligible low-income workers
Finance scheme	4. Service enhancement program	Capacity improvement through the construction of rail lines and provision of bus fleets, and service quality enhancement such as frequency and punctuality
	5. New finance scheme	Operating assets owned by the government, private operators bid to run services, government keeps the fare revenue and pays operators

Here we provide an example of how the theory of change analysis is applied to the distance-based fare scheme (Fig. 4). The distance-based fare scheme was introduced in 2010 and designed to replace a flat-rate fare scheme. Singapore's transport is a hub-and-spoke design, which is an efficient model to bring commuters to transport hubs but also incurs frequent transfers. Under a flat-rate fare scheme, every transfer incurs a cost that lowers the utility of public transport. However, technological advancements in automatic fare collection and Global Positioning System devices enabled the conditions to implement a distance-based fare scheme without incurring high capital expenditure investments.

Following the theory of change's intervention logic, the outcomes and effects of a policy intervention are listed horizontally. Secondary data sources describe how the distance-based fare scheme improved connectivity and integration of the public transport system (Public Transport Council Annual Report 2017/2018). Commuters were able to pay only for the total distance traveled and were not burdened by the number of transfers required to reach their destination. The scheme simultaneously removed the need to manage and calculate charges for bus–metro or bus–bus transfers, making transfers more seamless and convenient. As the immediate output of the scheme, according to the

Singapore Land Transport Statistics, a change in the average public transport fare from 0.98S$ (as of 2010 before implementation) to 0.92S$ (as of 2011 after implementation) was observed. Although public transport on average became more affordable with the lowered fare level, a decrease in public transport operator (PTO) profitability and the unequal distribution of profits among train and bus operators were also reported. In the long-term, the increasing usage of public transport is observed, according to the Land Transport Statistics Report published by the Land Transport Authority in Singapore.

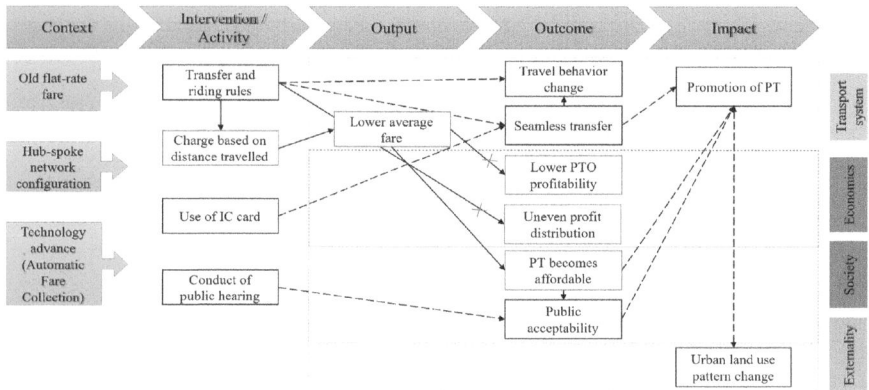

Fig. 4. Theory of change analysis of the distance-based fare scheme.

In this manner, the theory of change analysis generates a structural diagram of the links between interventions and observations. Multiple outcomes may be attributed to one intervention, whereas multiple interventions may contribute to one result. The solid and dotted line in Fig. 4 indicate which linkages are supported with quantitative data – such as the change in the average fare and the profits of the operators – and which are supported with qualitative observations. Negative unintended outcomes are also included and represented by red crosses over linkages.

The diagram provides a selective summary of the evaluation by extracting and describing observations that are highly relevant to the pricing dilemma. In summary, the distance-based fare scheme in Singapore improved public transport service and promoted the use of public transport. The impact of the scheme on the pricing dilemma was a lower average fare, thus improving availability; however, operator profitability also decreased especially for bus operators.

The same approach was applied to provide structured evaluations of the (i) new capacity factor fare adjustment scheme, (ii) workfare transport concession scheme, (iii) service enhancement program, and (iv) new finance scheme. The five analyses were then combined and integrated to produce a realist evaluation of the policy package, as shown in Fig. 5.

Balancing Affordability and Profitability in Urban Public Transport 157

The integrated realist evaluation is a straightforward process that combines the results. In the diagram, boxes representing context, scheme, intervention, and outcome are inherited from the theory of change analysis with individual pricing schemes differentiated by color. Only the "key factors" in the orange frames in the theory of change analysis (that are highly relevant to the pricing dilemma) are retained. Links between blocks are imported as well, including unintended negative outcomes. The interventions and outcomes are connected by mechanisms that can describe how the interventions bring about the outcomes, which should be simple in logic and evident in reasoning.

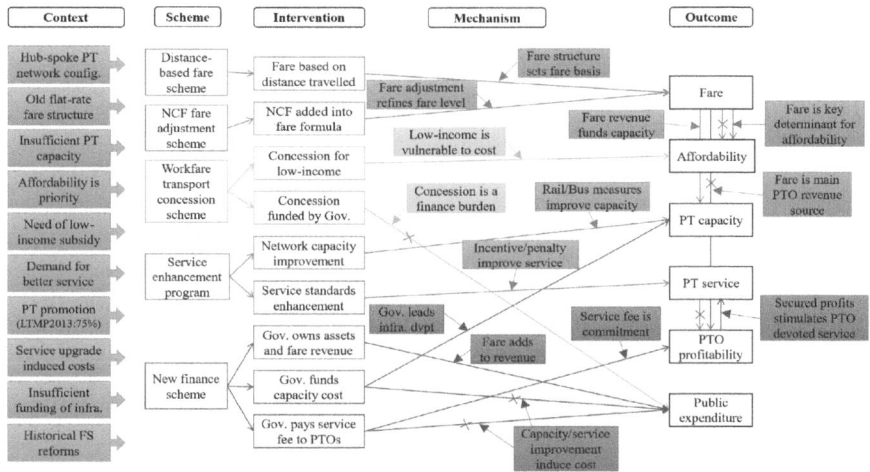

Fig. 5. Realist evaluation of Singapore's pricing policy.

3.2 Evaluating the Pricing Policy of Hong Kong

The same approach was applied to five of Hong Kong's pricing schemes, as summarized in Table 3 and 4.

Table 4 summarizes the CMOs of Singapore, focusing on the key factors impacting the pricing dilemma. For instance, the "Flat-rate" line in Table 4 is a summary of the first flow of the diagram in Fig. 5, highlighting how the fare scheme functioned in context to produce outcomes.

The generative causation process identifies and refines the sets of CMOs through triangulation with observations within the secondary data, eventually becoming independent of specific cases and generally applicable. In the next section, we present these refined causal relationships in the form of a causality map, illustrating the interactions between the pricing schemes and the pricing dilemma.

Table 3. Public transport pricing policy of Hong Kong.

Type	Pricing scheme	Brief description
Fare scheme	1. Fare adjustment scheme	Fare is adjusted based on the formula that considers the consumer price index, wage index, and productivity factor
Subsidy scheme	2. Work incentive subsidy scheme	To provide a subsidy to eligible low-income workers (mean-tested, up to HK$600 per month)
	3. Public transport fare subsidy scheme	To provide a subsidy to all commuters up to 25% of actual PT expenses of more than HK$400 (non-mean tested)
Finance scheme	4. Rail development strategy	Capacity improvement through the construction of rail lines and service quality enhancement such as frequency and punctuality
	5. Bus route rationalization scheme	Adjust bus routes in accordance with the demand and capacity changes

4 A Framework for Public Transport Pricing: A Causality Map of Pricing Strategies and the Pricing Dilemma

A causality map is an intuitive and efficient way of presenting the dynamic relationships between interventions and outcomes and is highly suitable for ex-post evaluations on transport policy [15]. By generalizing the structure of the theory of change and realist evaluations of pricing schemes, causality maps for Singapore and Hong Kong can be generated (see Fig. 6). The links begin with the pricing schemes and impact specific system elements positively or negatively (−). The placing of the system elements and pricing schemes are adjusted to avoid crossing links. The causality map guides the comparative discussion that is presented next.

Subsidies and Fare Affordability. Both Singapore and Hong Kong have implemented schemes for fare setting and adjustment. In Singapore, the fare structure shifted from a flat-rate to a distance-based fare scheme, which lowered the average fare level. A new capacity factor was added to the fare adjustment formula, which aims to better reflect the costs induced by public transport capacity improvements. In Hong Kong, the operators were authorized to set and adjust fares according to socioeconomic factors as well as in response to financial performance. The fare adjustment formula incorporated the CCPI, the nominal wage index, and a pre-determined productivity factor, among others.

In addition, incentive and penalty mechanisms were added to the fare schemes to balance the tradeoff between public affordability and operator profitability. In Singapore, "cap and collar", "profit sharing", and other carrot-and-stick mechanisms were implemented, and the operator's rate of return was held within a certain range. Hong Kong's "affordability-cap" set the ceiling for the fare increase to be no more than the

Table 4. Realist evaluation of CMO pricing strategies for Singapore.

Decision/dilemma context	Mechanism	Outcome
Flat-rate fare structure contributes to low affordability and profitability	Distance-based fare setting, and fare adjustment determines the fare level	Integrated fare promotes public transport
Social need for an affordable fare, especially for low-income households	Mean-tested concession and group-differentiated fare adjustment benefit low-income households	Affordability is maintained while implementing fare adjustments
Public transport network capacity is not sufficient for maintaining profitability/affordability	Construction of railway infrastructure and adding bus fleets increases network capacity	Capacity improved in both the short-term (bus) and long-term (rail), leading to increased costs
Demand for better public transport service while maintaining profitability/affordability	Incentive and penalty imposed on PTOs through service regulations	Service level and public satisfaction improved
Affordability priority constrains fare box revenue to operators	Service fee is paid by the government through finance structure reforms	PTO's profitability is secured while providing better services and keeping affordable fares
Capacity/service upgrading induce costs; fare revenue cannot meet investment gap, leading to insufficient profits	Government owns the assets, funds infrastructure, collects fare revenue, and pays costs	At the cost of public funds, PTOs' profitability is assured, and capacity upgrades are met

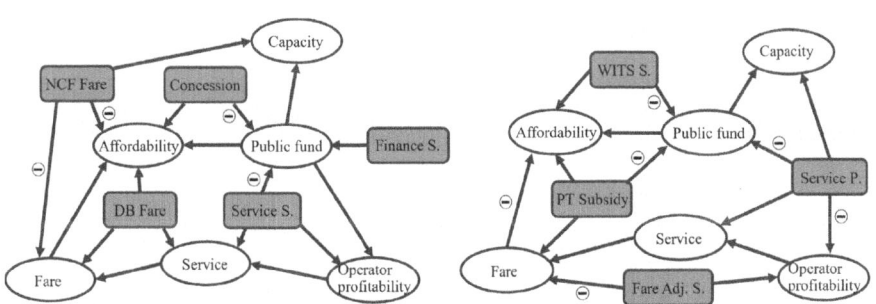

Fig. 6. Causality maps of Singapore (left) and Hong Kong (right).

year-on-year change in the household income for the previous year to cope with the ever-increasing fares.

However, both cases also show how the fare scheme alone is not sufficient in addressing affordability. Although differentiating social groups in fare adjustments could limit the rate of fare increase, it can still result in a fare increase for low-income groups. The subsidy scheme is a more straightforward, affordability-oriented instrument. Three subsidy schemes have been implemented in Singapore and Hong Kong, including the mean-tested Workfare Transport Concession Scheme, the Working Incentive Transport Subsidy Scheme, and the non-mean-tested Public Transport Subsidy Scheme – all backed by public funds.

Hong Kong's new Public Transport Subsidy Scheme, which is the first non-mean-tested subsidy in the world, shows that while the subsidy scheme incurred enormous costs for public funds (approximately $2.3 billion per year), the high-income groups would receive more subsidies than low-income groups (as high-income groups incurred higher expenses on transportation). Such results indicate the need to address structural deficiencies that can impact the distribution of benefits of a non-mean-tested subsidy.

Overall, Singapore and Hong Kong have presented different rationales in the management of fare affordability, as highlighted in the causality maps in Fig. 7. The Singaporean Government attempts to prevent the fare increases for users and bears the associated cost with the public fund, whereas Hong Kong allows the fare increase first but provides subsidies to the public through a public fund.

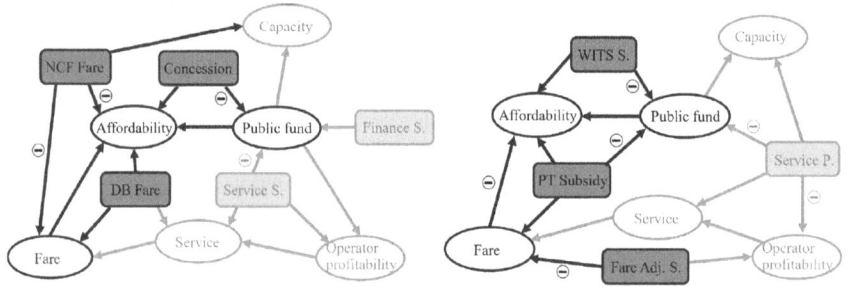

Fig. 7. Affordability in Singapore (left) and Hong Kong (right).

Finance Structure. Due to the relatively low fares that were maintained in Singapore due to pricing schemes, fare revenue was not sufficient to finance the construction of new lines and to maintain reasonable profitability for the operators. Due to the high costs associated with service enhancements, insufficient fare box revenues will prevent operators from maintaining high service levels. With the New Finance Scheme, infrastructure facilities and operating assets are placed in the hands of the government, although daily operations are still managed by the operators. This change of asset ownership can free operators from financial concerns and enable them to focus on maximizing the service enhancement of bus and rail services. Service quality is maintained by contractual obligations negotiated with the government. This arrangement also enables the government to undertake integrated and long-term planning for the entire public transport network.

The general context in Hong Kong is that the public transport sector historically emphasized the commercial principle. There is no direct link between public funds and operator profitability (as shown on the right side of Fig. 8). Instead, operator profitability is linked only with fare revenue via the fare scheme. While operator profitability is emphasized in pricing policy making, the intervention of the government in operator

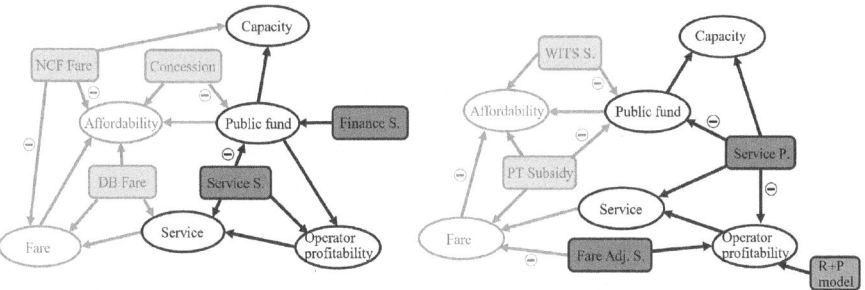

Fig. 8. Finance structure in Singapore (left) and Hong Kong (right).

revenue is constrained. The public transport operators do not receive any direct subsidy from the government but are authorized to collect fare revenue and mobilize alternative financing sources.

External financial sources are therefore important for maintaining the profitability of public transport operators. For example, through the Rail + Property model, the MTR Company in Hong Kong internalizes alternative sources of revenue by managing the land property and the associated income from real estate and sales. By doing so, the profits from property value increases can also compensate for infrastructure construction costs. Such financial structure makes Hong Kong one of the few cases in the world whose public transport operator is profitable without receiving any government subsidy.

5 Conclusion

This paper has presented an ex-post evaluation of the public transport pricing policy packages of Singapore and Hong Kong, to demonstrate the dynamic impacts of pricing schemes on the pricing dilemma. Both cases reveal how fare schemes, subsidy schemes, and finance schemes can affect the pricing dilemma. By integrating theory of change evaluations with realist evaluations, a generalized causality map can be derived which can be tested against other cases. Fare schemes, including adjustment schemes, can incorporate social and economic conditions such as inflation indicators and wages indices into formulas, thus reflecting the changes in income and expenses as well as the affordability of public transport. However, low-income groups can remain vulnerable to fare adjustments, and thus subsidy schemes that directly target the affordability of low-income groups can be more effective in maintaining affordability for these users. In general, a mean-tested subsidy could be more efficient in addressing affordability but complicated in terms of implementation than a non-mean-tested subsidy.

Finance schemes can set the financial structure of the public transport system, thus profoundly affecting the balance of the dilemma. The finance structure can be proactively reformed to cope with demand increases and investment needs. When we take a step back, however, we can observe that the public transport sector in Singapore has gone from nationalization, privatization, and back to nationalization in the past decades. For a private-oriented finance structure, the private sector could wield more power in the formulation of fare and subsidy schemes.

The integrated evaluation of each scheme into a realist evaluation allows us to understand both cases as a holistic policy package. Generalized mechanisms can be incorporated into the schemes to address the pricing dilemma incrementally, rather than as a one-shot solution.

This study provides a comprehensive evaluation of a series of pricing schemes including fare setting and adjustment, subsidies, and finance structure, on their impacts on the public transport pricing dilemma. Although this study is limited to Singapore and Hong Kong cases and to the pre–coronavirus 2019 period, interested readers and potential policy-makers should aim to test, validate, refine, or falsify the causal mechanisms of the general framework by exploring a wider range of existing urban public transport pricing cases. These findings should prove to be useful in the development and implementation of integrated pricing strategies for public transport policy-makers and operators.

Appendix: List of Data Source

There are four types of data sources:

- Report: Data collected through in-depth studies
- Legislation: Information from written documents
- Statistics: Quantitative information compiled by government institutions
- Survey: Information gathered through general public

Detailed data sources are listed below, sorted by cities.

Source	Singapore
Report	Looi (2007). Striking A Fare Deal: Singapore's Experience in Introducing a Fare Review Mechanism Li, M. (2018). International Benchmarking Study of Public Transportation Affordable fares, sustainable public transport: The Fare Review Mechanism Committee Report Leong. (2016). Improving bus service reliability: The Singapore experience Land Transport Authority. (1996). A World Class Land Transport System Land Transport Authority. (2018). Land Transport Authority Annual Report 2017/18 Land Transport Authority. (2019). LAND TRANSPORT MASTER PLAN (LTMP) 2040 Land Transport Authority. (2019). Performance of Rail Service Reliability
Legislation	Ministry of Transport of Singapore. (2018). Fares & Payment Systems Public Transport Council. (2018). 2018 Fare Review Exercise Public Transport Council. (2018). Commencement of 2018 fare review exercise Public Transport Council. (2018). Public Transport Council Annual Report 2017/2018 Ministry of Transport of Singapore. (2015). Workfare Transport Concession Scheme Land Transport Authority. (2013). LTA Masterplan 2013 Land Transport Authority. (2018). Financial Statements of Land Transport Authority

(*continued*)

(*continued*)

Source	Singapore
Statistics	Land Transport Authority. (2011). Singapore land transport statistics in brief Statistics Department of Singapore. (2019). Key Household Income Trends 2018 SMRT Company of Singapore. (2016). Briefing to Analysts on the New Rail Financing Framework One Singapore Organization. (2016). Welfare Schemes in Singapore
Survey	Land Transport Authority. (2012). Public transport customer satisfaction survey 2012 Public Transport Council. (2018d). Public Transport Customer Satisfaction Survey

Source	Hong Kong
Report	Transport and Housing Bureau. (2014). Discussion Paper on Railway Development Strategy 2014 Transport and Housing Bureau. (2015). Public Transport Topical Study on Franchised Bus Service Transport and Housing Bureau. (2018a). Discussion Paper for Finance Committee on Public Transport Subsidy Scheme Yeung, S. (2012). The study of transport subsidies in Hong Kong Transport Advisory Committee. (2014). Report on Study of Road Traffic Congestion in Hong Kong Transport and Housing Bureau. (2017). Public Transport Strategy Study Final Report National Surface Transportation Infrastructure Financing Commission. (2009). Paying Our Way: A New Framework for Transportation Finance
Legislation	Legislative Council Panel on Transport. (2019). MTR Fare Adjustment for 2019 Legislative Council Panel on Transport. (2018). Implementation of the Public Transport Fare Subsidy Scheme Government Budget. (2019). Head 186 - Transport Department Hong Kong Transport Department. (2009). Review of Fare Adjustment Arrangement for Franchised Buses Hong Kong Transport Department. (2019). Legislative Council Panel on Transport Fare Increase Applications Transport and Housing Bureau. (2018). Public Transport Fare Subsidy Scheme Labor Department. (2017). Proposed freezing of income limits under Work Incentive Transport Subsidy Scheme Labor Department. (2018). Work Incentive Transport Subsidy Scheme Legislative Council. (2017). Background Brief on Work Incentive Transport Subsidy Scheme

(*continued*)

Source	Hong Kong
Statistics	Statistics Section of Transport Department. (2019). Monthly Traffic and Transport Digest 2019 Working Family Allowance Office. (2019). Examination of Estimates of Expenditure 2019-20 Environmental Protection Department. (2017). Air Quality in Hong Kong 2017 Government Budget. (2019). Working family and student financial assistance agency
Survey	HKSAR Transport and Housing Bureau. (2016). Public Consultation on the Review of the MTR Fare Adjustment Mechanism

References

1. Chua, V.C.H.: Comparison of Rail Fares Between Singapore and 35 Major Cities Around the World. Public Transport Council, Singapore (2016)
2. Mijares, A.C., Regmi, M.B.: Enhancing the sustainability and inclusiveness of the Metro Manila's urban transportation systems: proposed fare and policy reforms. Transp. Commun. Bull. Asia Pac. **84**, 28–40 (2014)
3. Serebrisky, T., Gómez-Lobo, A., Estupiñán, N., Muñoz-Raskin, R.: Affordability and subsidies in public urban transport: what do we mean, what can be done? Transp. Rev. **29**(6), 715–739 (2009). https://doi.org/10.1080/01441640902786415
4. Sayyadi, R., Awasthi, A.: A system dynamics based simulation model to evaluate regulatory policies for sustainable transportation planning. Int. J. Model. Simul. **37**(1), 25–35 (2017). https://doi.org/10.1080/02286203.2016.1219806
5. Worsley, T.: Ex-post Assessment of Transport Investments and Policy Interventions (ITF Roundtable Reports). International Transport Forum. OECD (2017)
6. OECD: OECD Regulatory Policy Outlook 2015. OECD (2015)
7. Eastwood, J.G., et al.: Implementation, mechanisms of effect and context of an integrated care intervention for vulnerable families in central Sydney, Australia: a research and evaluation protocol. Int. J. Integr. Care **19**(3), 1–13 (2019). https://doi.org/10.5334/ijic.4217
8. Hills, D., Junge, K.: Guidance for transport impact evaluations - choosing an evaluation approach to achieve better attribution. Tavistock Institute (2010)
9. UNICEF: The South African Child Support Grant impact assessment: evidence from a survey of children, adolescents and their households (2014)
10. Pawson, R.: Realistic Evaluation. SAGE Publications Ltd., London (1997)
11. Blamey, A., Mackenzie, M.: Theories of change and realistic evaluation: peas in a pod or apples and oranges? Evaluation **13**(4), 439–455 (2007). https://doi.org/10.1177/1356389007082129
12. Litman, T.: Developing indicators for comprehensive and sustainable transport planning. Transp. Res. Rec. **1**, 10–15 (2017). https://doi.org/10.3141/2017-02
13. Jeon, C.M., Amekudzi, A.A., Guensler, R.L.: Sustainability assessment at the transportation planning level: performance measures and indexes. Transp. Policy **25**, 10–21 (2013). https://doi.org/10.1016/j.tranpol.2012.10.004
14. Litman, T., Burwell, D.: Issues in sustainable transportation. Int. J. Glob. Environ. Issues **6**(4), 331–347 (2006). https://doi.org/10.1504/IJGENVI.2006.010889
15. Graham, D.: Causal Influence for Ex-post Evaluation of Transport Interventions. International Transport Forum. OECD (2014)

A Group Support System for Resolving Variation Disputes in Tunnel Construction Projects

Muhammad Tajammal Khan[✉] and Masahide Horita

Department of Civil Engineering, The University of Tokyo, 7-3-1 Hongo, Bunkyo-ku, Tokyo, Japan
{tajammal,horita}@g.ecc.u-tokyo.ac.jp

Abstract. Tunnel projects frequently face variations, delays, and disputes that are difficult to settle with traditional project management techniques due to fragmented documentation and conflicting goals of contracting parties. The cost and time required for dispute handling necessitate advancements in management skills for efficient resolution. This study proposes a group support system (GSS), which incorporates the building information modeling (BIM) and Nash bargaining principles to facilitate data-driven, transparent, and fair dispute resolution. The BIM framework identifies the modified scope and predicts the potential outcome of variation claims from the actual project dataset, while the game theory finds the optimal settlement by balancing the employer's and contractor's payoffs to formalize the negotiation strategy. A case study on a variation dispute from a tunnel project demonstrates that the engineer rejected the claim at the first instance, and the contractor submitted an exaggerated claim for time extension. The proposed GSS predicted the dispute outcome based on construction dataset and utility functions of parties, demonstrating that it can enhance negotiation and dispute resolution efficiency for fair settlements. Finally, the research advocates for the digital transformation in the construction industry for claim management, negotiations, and dispute resolution to improve transparency, efficiency, and collaboration.

Keywords: Variation Disputes · Building Information Model · Game Theory · Group Support System

1 Introduction

The increasing complexities in construction projects and deviation of physical condition from design cause variations and construction disputes [1]. Variations interrupt the construction workflow [2] through redesign [3], rework [4], and material reordering [3]. Variations not only undermine the project outcome in terms of cost overruns and delays [2] but are also a primary factor in costly and time-consuming claims, negotiations, and disputes [5, 6].

According to the global construction disputes report, the average value of disputes in the construction industry has increased to US$ 54.26 million, and disputes consume an average time of 13.4 months [7]. Traditional project management relies on two-dimensional (2D) drawings, three-dimensional (3D) models [5, 8], and fragmented paper-based documentation [9], which makes it difficult to retrieve accurate information required as evidence to manage and resolve the disputes [10]. Planning and scheduling software, such as Microsoft Project and Primavera [11], lack a dynamic link between drawings, construction workflow, and schedules [12], which limits their effectiveness in managing variations and resolving related disputes. Improving existing project management skills is crucial to resolving costly and time-consuming disputes through the adoption of technological developments.

Advancements in artificial intelligence and group support systems (GSSs) can potentially reduce time and effort by streamlining the decision process and enhancing collaboration [13, 14]. Various support systems are developed to analyze and predict the potential outcome of disputes [15, 16] for asset management [17], natural gas distribution planning [18], family law [19], offshore oil exploration [20], supply chain [21], and hiring subcontractors in the construction industry [22]. GSSs incorporate various methodologies, including decision tables, multi-criteria decision analysis, graph model for conflict resolution, and game theory to facilitate negotiation and predict potential outcomes of disputes, thereby enabling contracting parties to make informed decisions [23].

Among these methodologies, game theory provides a structured approach for resolving conflicts between stakeholders by modeling their strategic interaction and incentives [24, 25]. In particular, the cooperative game model [26] ensures mutually beneficial outcomes among contracting parties [27, 28] to find optimal outcomes to maximize joint payoffs while addressing the individual payoffs [25, 29]. Despite these advantages, the construction industry remains behind in adopting digital technologies for dispute resolution, highlighting the need for an advanced GSS for variation dispute resolution.

Building Information Modelling (BIM) is transforming project management by enhancing visualization, improving collaboration, and information management. Existing frameworks in BIM are focused on checking design completeness and clash detection [30, 31], avoiding variations [6, 32], variation propagation or ripple effect [5, 33], and dispute avoidance and management [6, 9, 34, 35]. BIM frameworks are proposed based on typical design workflows and lack the required flexibility to handle the dynamic and evolving nature of construction, as variations have the potential to disrupt progress, change construction workflow, and even halt the construction activities. Therefore, existing frameworks are unable to provide the required support for resolving disputes.

Given the limitation of traditional project management in resolving the variation disputes, this study aims to formulate GSS for dispute resolution. GSS integrates BIM for three-dimensional (3D)/four-dimensional (4D) modeling of tunnel and claim to store large project construction data. Artificial intelligence analyzes the large datasets to identify the scope of variation and predict the potential outcome of disputes. The 4D BIM model provides support to visualize the construction methodology and identify any change in workflow compared to the planned methodology. In this way, the proposed GSS will offer data-based support to contracting parties in transparent and efficient dispute settlement. GSS also uses Nash bargaining principles to estimate the optimal negotiation settlement for the dispute between the parties by balancing the payoffs and utilities to formulate the negotiation strategy.

Finally, the study demonstrates that BIM and game theory within GSS can help parties to formulate negotiation strategies that enhance negotiation efficiency and improve decision-making for dispute resolution.

2 Literature Review

Major causes of disputes in construction projects include variation (change order), design error, payment problems, site problems, contractual problems, and opportunistic behavior [6]. Variation is initiated by the employer for any change to the quantities of work, change to quality, change to physical parameters of any component, additional work, omission of a component, and change to the sequence or timing of the execution [36]. Inadequate site investigations [30, 37], design errors, or defects resulting from the designer's unfamiliarity with the site conditions and lack of experience [38] can all lead to variations. Most variations necessitate additional costs and time for completion, often leading to claims and disputes.

The contradictory goals of the contracting parties also cause disputes as parties get involved in contractual confrontations rather than solving the conflict at the initial stage [39]. The design of projects tends to be overly optimistic, resulting in inadequate funding for unforeseen physical contingencies. Public authorities are hesitant to decide the variations and accumulate for late settlements in courts [40]. The contractor demands the expeditious approval of the variation, as any delay or non-approval can result in the withholding of the contractor's due payment, thereby creating a cash flow issue. The contractor also tries to maximize the profit for variation without competing with any other contractor for work [41]. These causes lead to complex disputes in the construction industry, consuming significant time, finances, and human resources and necessitating an efficient dispute resolution GSS incorporating advanced technologies.

GSSs have been developed since the 1970s, primarily to enhance dispute resolution efficiency by predicting conflict outcomes and assisting the parties to make decisions based on accurate information. GSSs serve as intermediaries facilitating the parties to reduce negotiation time and cost [15, 16] by providing IT support to ensure that human negotiators control the decision-making process [42]. GSSs incorporate methodologies like multi-criteria decision analysis, graph model for conflict resolution, game theory, and statistical methods to predict the dispute outcome [23]. GSS helps both parties to be more effective and fairer, particularly in complex and strategically important disputes [14].

Among the methodologies, game theory provides tools for understanding the cooperative and competitive behavior of stakeholders for a fair and optimal resolution of conflicts [25]. The Nash bargaining solution [26] is particularly helpful in dispute resolution as it offers to balance individual payoffs while maximizing the combined payoffs, ensuring transparency and fairness in the process [27, 28]. This model is also explored for delay analysis and fair cost allocation among contracting parties [25, 29]. On the other hand, the construction industry adopts traditional management skills and is lagging in the operationalization of this theoretical concept in automated and digital environments [43].

Traditional project management utilizes complicated bar charts and critical path methods [44], making retrieval of documentary evidence challenging and potentially jeopardizing the entire negotiation process in complex disputes in the construction industry [45]. The unclear scope of varied works and inaccurate recording of verbal instructions are primary sources of disputes [46].

To improve the traditional management approaches, extensive research has been carried out in the past decade regarding the benefits of adopting BIM in the construction industry. The most adopted BIM benefits include accurate cost estimation, enhanced planning and scheduling, an enhanced collaborative environment, and improved visual management [5, 6, 47]. Researchers have examined the impact of these benefits to optimize design [48], improve project schedule [12, 32], foresee future impact of variations [5], understand the ripple effect of an intended variation on project schedule [4, 5], potential claims of variation [33], feasibility for claim management [9], and improve collaboration between project stakeholders [12].

While the integration of schedule (time) within 4D simulation can provide significant advantages for visualization of planned and as-built construction workflow, research on adoption of BIM for dispute resolution remains limited. Existing BIM frameworks primarily focus on design and schedule optimization. These frameworks do not support the dynamics of negotiating and settling construction disputes. Given the complexity of construction disputes, comprehensive GSS incorporating 4D BIM could significantly improve dispute resolution, particularly arising from variations and conflicting goals among parties.

3 Methodology for BIM Framework

Figure 1 provides a brief overview of the proposed BIM framework developed based on design and construction data to find the potential outcome of variation disputes. The planned and as-built BIM models were created using Autodesk Revit phases and Navisworks Manage, two popular BIM programs in the construction industry. A detailed work-breakdown-structure (WBS) of each component can be defined to input information regarding unique element identity, tunnel reach (location), physical parameters like length, area, and volume, and time (schedule), thus creating a large dataset about the project. Component information from 3D models is exported to Excel sheets, and a developed Python program analyzes the exported large dataset to identify the varied scope of works. The Python program carries out the component-by-component comparison of planned and as-built models by matching unique element identity and related physical parameters to identify the modified, omitted, and additional scope of work. The program can also extract the planned and as-built duration to complete each component essential to estimate the outcome of claim.

Similarly, the unique schedule identity is defined for each component to connect the time (schedule), which makes it easier to simulate the construction workflow in 4D in Navisworks. The project parties can visualize any change in the construction workflow during the execution of the project by comparing it with the planned workflow. This will allow human negotiators to decide the negotiation strategy and take decisions based on correct information in a fair, transparent, and timely manner. This information-based negotiation will lower the transaction costs for prolonged variation claims and disputes.

This generalized methodology of the proposed BIM framework can be applied to a wide range of infrastructure projects like buildings and bridges. While Autodesk Revit

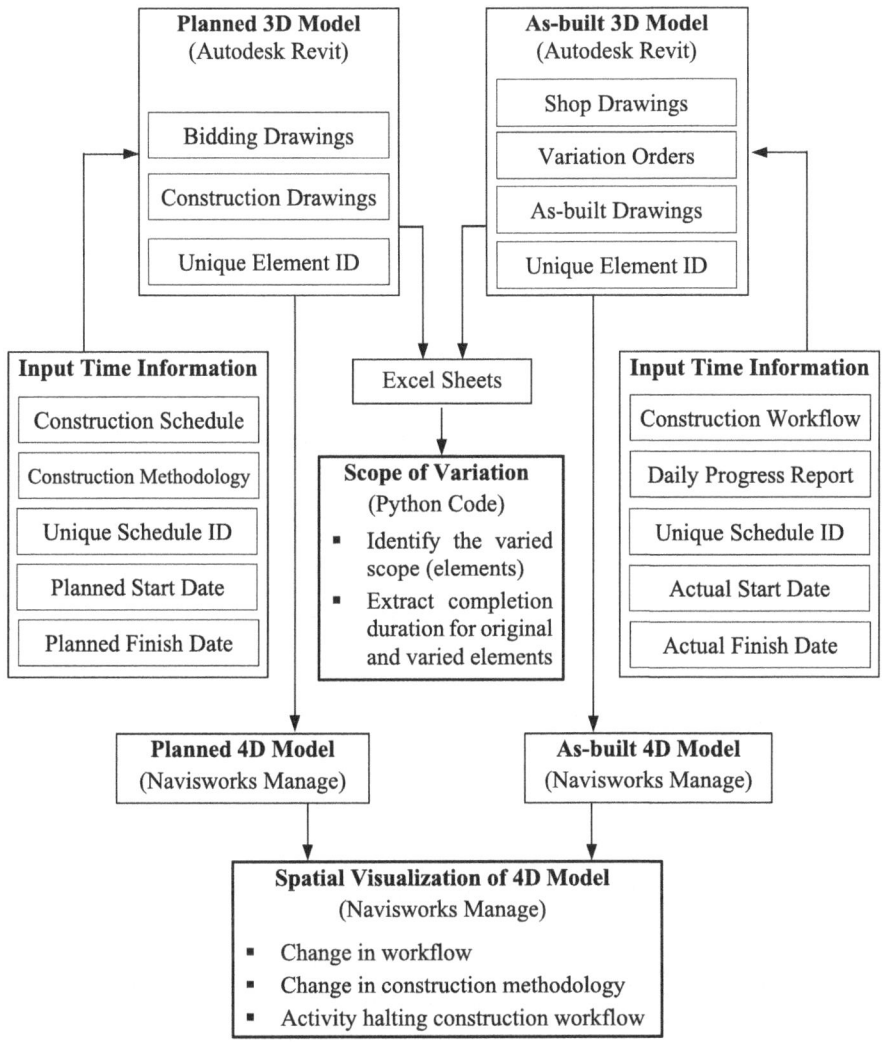

Fig. 1. Proposed BIM framework for predicting potential outcome

provides predefined elements like columns, beams, slabs, and foundations, making it a suitable tool for modeling buildings and bridges. The modeling of underground tunnels in Autodesk Revit presents significant challenges due to irregular geometry and construction dynamics. Prior studies by Moayeri et al. [4] and Handayani et al. [5] have developed BIM frameworks tailored to building projects to identify variation in scope. The proposed framework and methodology in this study can be adopted for other infrastructure projects by defining the unique element identities and incorporating additional data related to construction schedule, construction methodologies, and progress reports.

4 Case Study of Hydropower Project

The proposed BIM framework is used to predict the potential outcome of an ongoing variation dispute in an anonymous hydropower project [49]. The project comprises three access tunnels: (i) the main access tunnel (MAT), (ii) the temporary access tunnel (TAT), and (iii) the access tunnel to the gate chamber (TGC), and two diversion tunnels (DT-1 and DT-2). The construction of TAT aimed to facilitate intermediate access to DT-1 and DT-2, thereby speeding up the construction of diversion tunnels. The two diversion tunnels will divert the river water for the construction of the main dam and other ancillary components.

FIDIC-MDB Harmonized Edition (2010), published by the International Federation of Consulting Engineers (FIDIC), is signed as a contract document between the parties [36, 49]. The employer transferred subsurface risks to the contractor, adopting an extremely conservative approach. Deviation in rock formations along the tunnel and installation of additional rock support will not allow the contractor to claim extra time for completion.

Soon after the commencement of construction works, the project activities were badly affected by the global spread of COVID-19. Access tunnels and diversion tunnels have been completed after a delay of approximately 3 years due to various contractual issues, including a modification in the design at the junction of TAT and DTs.

4.1 Claim Background

The engineer modified the drawings for the junction of TAT with DT-1 and DT-2 by increasing the size of TAT as shown in Fig. 2, and proposing additional rock support. The design was modified for reach TAT $0 + 070.4$–$0 + 110.4$ (junction with DT-1) and TAT $0 + 145.5$–$0 + 165.43$ (junction with DT-2) as shown in Fig. 3. The change in design is a variation under sub-clause 13 [Variations and Adjustments] of the contract, while the provision of additional rock support does not constitute a variation under the technical specifications. The contractor submitted a claim for a 30-day extension of time based on the installation of additional rock support under sub-clauses 8.4 [Extension of Time for Completion], 13 [Variations and Adjustments], and 20.1 [Contractor's Claim] without highlighting the increase in size of TAT. The engineer rejected the variation claim, resulting in a dispute that has remained unresolved for approximately 3 years. The contractor has submitted the claim alongside a separate COVID-19 claim, which lowers the contractor's efficiency (production rates) in tunnel completion.

4.2 Claim Analysis in BIM Framework

Planned and as-built BIM models are prepared based on information extracted from project drawings, construction schedules, and progress reports utilizing the proposed BIM framework (Fig. 1). Models are enriched with information related to component identity, physical parameters, and schedule (time). Following the methodology of the framework, the modified tunnel reaches of the TAT are identified, including the extraction of time information. Details of identified tunnel reaches representing variation are outlined in Table 1.

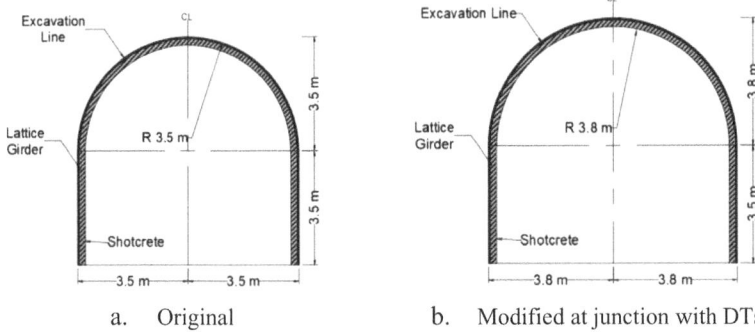

Fig. 2. Typical excavation section of TAT

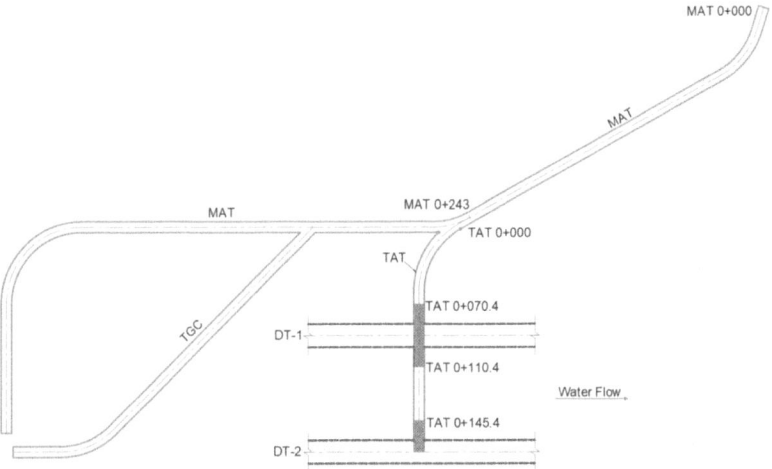

Fig. 3. Layout plan of underground tunnels and reaches of varied scope of works

Comparing the identified varied scope of works with the drawing (Fig. 3) reveals that the proposed BIM framework has accurately identified the components and reaches of modified TAT. In addition, the framework also predicted the planned and as-built duration to complete the modified and/or varied scope of TAT (Table 1). The contractor originally planned 24 days for completion of the identified tunnel reaches as per the baseline schedule, while the contractor completed the works in 31 days; thereby, additional scope was constructed in 7 days. Predicted 7 days also include the impact of COVID-19, which has also been claimed separately for overall project completion. The prediction of extra time for completing the variation can aid human negotiators to take efficient decisions about the dispute.

While the proposed BIM framework predicts a maximum 7-day extension for completion of varied scope from project construction data, the employer and contractor have conflicting goals in accepting this outcome. To formalize the negotiation strategy, Nash

Table 1. Varied scope of works identified by proposed BIM framework

Sr #	Tunnel Reach	Element ID	Planned (days)	As-built (days)	Remarks
1	TAT (0 + 070.5–0 + 073.43)	TAT_1.26	1	4	TAT junction with DT-1
2	TAT (0 + 073.43–0 + 076.93)	TAT_1.27	1	4	
2	TAT (0 + 076.93–0 + 079.5)	TAT_1.28	1	1	
4	TAT (0 + 079.5–0 + 081.93)	TAT_1.29	1	2	
5	TAT (0 + 081.93–0 + 085.5)	TAT_1.30	1	1	
6	TAT (0 + 085.5–0 + 088.5)	TAT_1.31	1	1	
7	TAT (0 + 088.5–0 + 091.5)	TAT_1.32	1	1	
8	TAT (0 + 091.5–0 + 095)	TAT_1.33	2	2	
9	TAT (0 + 095–0 + 098.93)	TAT_1.34	2	2	
10	TAT (0 + 098.93–0 + 101.5)	TAT_1.35	1	2	
11	TAT (0 + 101.5–0 + 103.93)	TAT_1.36	1	1	
12	TAT (0 + 103.93–0 + 107.43)	TAT_1.37	2	1	
13	TAT (0 + 107.43–0 + 110.5)	TAT_1.38	1	1	
14	TAT (0 + 145.5–0 + 148.43)	TAT_1.51	1	1	TAT Junction with DT-2
15	TAT (0 + 148.43–0 + 150.5)	TAT_1.52	1	1	
16	TAT (0 + 150.5–0 + 151.93)	TAT_1.53	1	1	
17	TAT (0 + 151.93–0 + 154.43)	TAT_1.54	1	1	
18	TAT (0 + 154.43–0 + 156.93)	TAT_1.55	1	1	
19	TAT (0 + 156.93–0 + 160)	TAT_1.56	1	1	
20	TAT (0 + 160–0 + 163)	TAT_1.57	1	1	
21	TAT (0 + 163–0 + 165.43)	TAT_1.58	1	1	
	Total =		24	31	

bargaining principle is applied to find a fair and optimal settlement of claim considering the utility and payoff of the parties.

4.3 Nash Bargaining Solution and Payoff Analysis

The variation dispute is modeled assuming that the contracting parties seek mutually beneficial resolution rather than prolonging the dispute. This section uses Nash bargaining principles to theoretically analyze the dispute to formalize the negotiation strategy for determining the fair and optimal settlement based on utilities and conflicting behavior. In this model, $U_E(t)$ and $U_C(t)$ represent agreement utility for the employer and contractor respectively, while d_E and d_C are disagreement payoffs. The maximization problem for the utilities can be defined as:

$$Max(U_E(t) - d_E) \cdot (U_C(t) - d_C)$$

subject to:

$$t \in S, U_E(t) \geq d_E, U_C(t) \geq d_C \tag{1}$$

In the above equation, t represents the extension of time, S is the set of feasible solutions for the extension of time. The utility functions are formulated as follows:

$$U_E(t) = R - W_C - W_M - a(t - t_c) \tag{2}$$

$$U_C(t) = W_C - C_E + b \cdot t \tag{3}$$

where, R is expected revenue for the employer from completing the tunnels, W_C original payment to the contractor without an extension of time, W_M is the miscellaneous expenses by the employer (e.g. engineer and other project management), a is a factor converting the time into utility loss for the employer and t_c is maximum extension claim from the contractor, $a(t - t_c)$ is the cost to the employer for agreeing to an extension of time, C_E is the contractor's effort and/or direct cost associated with construction works, and $b \cdot t$ is additional benefit to the contractor for securing an extension of time.

If the negotiations fail and the dispute escalates to dispute boards, arbitration, and court proceedings, additional transaction costs related to dispute management and litigation costs, cash flow losses, potential contract termination, and reputational damage significantly impact the disagreement payoffs of the parties. Claim management and litigation costs are not static but increase proportionally with the duration of dispute, making disagreement payoffs a dynamic function. Therefore, prolonged duration of the dispute may substantially increase disagreement payoffs approaching the utilities of parties at some point, satisfying the function (1). While the cash flow losses to the contractor, reputational risks, and potential contract termination to the parties play a critical role in real-world dispute resolution, their qualitative costs are ignored for simplification of the mathematical model. The disagreement payoffs for the employer and contractor will be:

$$d_E = R - W_C - W_M + W_D \cdot t_d \tag{4}$$

$$d_C = W_C - C_E + C_D \cdot t_d \tag{5}$$

where W_D and C_D represent the dispute handling costs per day to the employer and contractor, respectively and t_d is the dispute duration between the contracting parties. Increase of these costs over time creates a financial burden for both the parties and compel to cooperate and timely resolve the dispute instead of prolonging it. The prolonged dispute will cumulatively increase these costs and may erode the benefits of holding out for a more favorable settlement, thereby making early negotiation for more rational strategy. Solving the maximization problem (1) provides the optimal settlement T_S time as:

$$T_S = \frac{abt_c + (aC_D - bW_D) \cdot t_d}{2ab} \tag{6}$$

This Nash bargaining solution is applied to find the optimal settlement and corresponding payoffs to parties from the claim theoretically. Based on project approval documents, the total expected revenue R from the diversion tunnels over a design life of 50 years is assumed to be US$ 1.5 billion. The original payment to the contractor W_C for the construction of tunnels is US$ 175 million, while additional expenses W_M for engineering and project management are assumed to be US$ 65 million. The contractor's cost of construction C_E is assumed to be 90% of W_C providing 10% profit from the signed contract.

The utility factor a is calculated as US$ 82,000 from the revenue loss to the employer, while the utility factor b is taken as US$ 90,000 from the contract documents against the delay in completion of diversion tunnels. t_c is 30-day extension of time claim as submitted by the contractor, t_d is the dispute duration and is assumed as 3-years from project data. The daily management cost of dispute is assumed to be US$ 100 and USD 150 for the contractor C_D and employer W_D, respectively. These costs do not include the litigation cost as the dispute is not yet referred to dispute board or arbitrators for decision, which may increase the transaction cost exceptionally. Using these inputs in the Nash bargaining equation, the optimal settlement of claim is 14.6 days. Table 2 summarizes the corresponding payoffs for the employer and contractor:

Table 2. Individual Payoff's for the Parties

Sr #	Methodology	Time (day)	U_E (US$)	U_C (US$)	d_E (US$)	d_C (US$)
1	Nash solution	14.6	1,261,262,800 (1,098,550)	18,814,000 (1,204,500)	1,260,164,250	17,609,500
2	BIM framework	7	1,261,886,000 (1,886,000)	18,130,000 (630,000)	1,260,164,250	17,609,500

Note: (-) values in bracket represents the increase in utility

Table 2 also presents the payoffs for BIM's predicted outcome of a 7-day time extension. In this case, the transaction cost for dispute management would be minimal and is assumed to be zero for the calculation of disagreement utilities for both parties. Both the solution, i.e., BIM and Nash principles, increase the payoff for the parties satisfying the boundary conditions and provide mutually beneficial outcomes to parties.

4.4 Parties Behavior and Discussion

The engineer modified the size of TAT in the construction drawings, which is a variation under the condition of the contract. However, the engineer did not issue variation instructions under the contract to avoid project delay. Additionally, the engineer rejected the contractor's notice of claim for time extension. Therefore, this rejection, or no extension of time under the claim, is the disagreement utility d_C for the contractor. In response to the rejection of the notice to claim, the contractor submitted an exaggerated claim of a 30-day extension of time for variation, which constitutes the disagreement utility for the employer d_E. The continuous exchange of correspondence is increasing the transaction cost and wasting the time and financial resources of the parties. However, the actual time required to complete varied work falls between the two extremes, which will constitute a fair and optimal settlement to the dispute.

The predicted outcome from the BIM framework and Nash bargaining solution represents two different perspectives on dispute negotiation and resolution. BIM framework incorporates the geometry and construction timelines of each tunnel element in the planned and as-built 3D/4D models. The framework analyzes the large project dataset to identify the scope of variation in tunnel project. Therefore, the BIM framework provides a technical and factual estimate of the extra time based on actual project construction data from the planned and as-built models. On the other hand, the Nash solution offers a negotiation-based settlement, considering the payoffs and strategic behaviors of both employer and contractor. The Nash bargaining approach incorporates stakeholder utilities, disagreement payoffs, and transaction costs associated with prolonged disputes. This approach maximizes joint payoffs by balancing employer revenue losses against the contractor's need for additional compensation due to variation in scope of work. The inclusion of the BIM framework and Nash principles in GSS provides practical and theoretical insight into the variation claim to devise the negotiation strategies as BIM outcome is data-driven and Nash solution is equating the payoff for the parties.

BIM minimizes information asymmetry and manipulation by finding the data-driven outcome of the claim as parties have the same dataset and dispute prediction. It also reduces the influence of power-driven negotiations towards the evidence-based negotiations. However, a sensitivity analysis can further elaborate on the power imbalance in negotiation by incorporating the BIM-based factual evidence and risk aversion of the contractor into the Nash bargaining solution. This will strengthen the GSS across various negotiation scenarios, enhancing its applicability in real-world dispute negotiations. The authors are expanding this aspect as the future scope of work under the research study.

5 Conclusion

This study advances dispute resolution in tunnel construction projects by proposing GSS that incorporates the BIM technology and Nash bargaining principles to optimize negotiation strategies. GSS bridges the gap between technical solutions from project data and defining structured strategies from game theory principles, enabling a fair, data-driven, and efficient negotiation process. The BIM framework analyzes the real-time project data and provides a fact-based outcome of variation claims, unlike the traditional negotiations that rely on fragmented documentation and experts' perception/opinion. Integration of Nash bargaining principles further strengthens the dispute settlement process by balancing the parties' utilities to achieve a fair and mutually beneficial settlement of variation disputes.

Power imbalance, information asymmetry, and parties' biases often influence traditional dispute resolution mechanisms, leading to prolonged conflicts and unfair settlements. BIM framework in GSS ensures that the parties share the same data, which minimizes the manipulation and misinterpretation of variation claims. Nash solution approach further enhances fairness in the negotiation process by quantifying the payoffs from the claim and optimizing settlement based on maximizing utilities.

Proposed GSS increases efficiency in variation dispute resolution in the construction industry. Traditional project management involves manual claim identification and data extraction, which is time-consuming and leads to prolonged claims, increasing handling and litigation costs. GSS automates data processing by reducing time for claim identification or verification and determining the potential claim outcome. It minimizes the time for claim settlement and the transaction cost associated with handling claims and litigation. A case study on a hydropower project demonstrates the effectiveness of GSS, wherein the BIM predicted a 7-day time extension for completion of varied scope in TAT, while Nash bargaining solution suggests a 14.6-day time extension balancing the contractor's compensation and employer's revenue loss. The two solutions can facilitate the parties to negotiate the dispute and settle in a transparent and fair manner.

The study advocates for technological advancements in claim management and resolution, supporting GSS as an innovative alternative to traditional project management skills. It combines BIM, Python AI, and game theory, demonstrating its potential to identify and estimate the possible outcome of the variation claim. Adoption of technology and data-driven solutions will provide a strong base for fair and efficient negotiation and resolution of variation disputes in construction projects.

6 Limitation

The proposed GSS has been validated through a variation claim on a real-world tunnel construction project. Its effectiveness for other types of infrastructure projects require further investigation and evidence. BIM framework is applicable to other infrastructure projects as explained in Sect. 3. Nevertheless, a study of variation claims from real-world large infrastructure project is essential to access the efficacy of the proposed BIM framework in resolving the variation claims.

Nash bargaining solution within GSS offers theoretical insight into the negotiation dynamics of negotiation and interaction between the parties. Higher claim handling and litigation expenses in other infrastructure projects, particularly in buildings, may significantly influence the decision-making, compelling stakeholders to seek early dispute resolution. In such projects, the tendency of disagreement payoffs to approach agreement utility escalates due to high litigation costs relative to lower compensation to contractor or lesser revenue loss for the employer from time extension due to variations.

7 Future Research Work

Future research will focus on performing a sensitivity analysis that integrates the BIM-based evidence and contractor's risk aversion into Nash bargaining solution. This will yield deeper understanding of parties' risk preferences and the influence of BIM evidence on negotiation strategies and claim settlement. Integrating risk and BIM evidence will further refine the GSS, improving its applicability and robustness in actual dispute negotiations.

Acknowledgments. This work was supported by JSPS KAKENHI Grant Number JP22H01561.

Disclosure of Interests. The authors have no competing interests to declare that are relevant to the content of this article.

References

1. Park, M.: Dynamic change management for fast-tracking construction projects. In: Proceedings 19th International Symposium on Automation and Robotics in Construction (ISARC), pp. 81–89. International Association for Automation and Robotics in Construction, London (2002)
2. Sun, M., Meng, X.: Taxonomy for change causes and effects in construction projects. Int. J. Project Manage. **27**(6), 560–572 (2009). https://doi.org/10.1016/j.ijproman.2008.10.005
3. Yap, J.B.H., Abdul-Rahman, H., Chen, W.: Collaborative model: managing design changes with reusable project experiences through project learning and effective communication. Int. J. Project Manage. **35**(7), 1253–1271 (2017). https://doi.org/10.1016/j.ijproman.2017.04.010
4. Moayeri, V., Moselhi, O., Zhu, Z.: Design change time ripple effect analysis using a BIM-based quantification model. In: Proceedings Construction Research Congress 2016, pp. 28–36. ASCE, Reston, VA (2016)
5. Handayani, T.N., Likhitruangsilp, V., Yabuki, N.: A building information modelling (BIM)-integrated system for evaluating the impact of change order. Eng. J. **23**(4), 67–90 (2019). https://doi.org/10.4186/ej.2019.23.4.67
6. Wang, J., Zhang, S., Fenn, P., Luo, X., Liu, Y., Zhao, L.: Adopting BIM to facilitate dispute management in the construction industry: a conceptual framework development. J. Constr. Eng. Manage. **149**(1), Article 03122010 (2023). https://doi.org/10.1061/(ASCE)CO.1943-7862.0002419
7. ARCADIS: Global construction disputes report: a road to early resolution (2021)
8. Mehrbod, S., Staub-French, S., Mahyar, N., Tory, M.: Characterizing interactions with BIM tools and artifacts in building design coordination meetings. Autom. Constr. **98**, 195–213 (2019). https://doi.org/10.1016/j.autcon.2018.10.025
9. Shahhosseini, V., Hajarolasvadi, H.: A conceptual framework for developing a BIM-enabled claim management system. Int. J. Constr. Manage. **21**(2), 208–222 (2021). https://doi.org/10.1080/15623599.2018.1512182
10. Haugen, T., Singh, A.: Dispute resolution strategy selection. J. Leg. Aff. Dispute Resolut. Eng. Constr. **7**(3), 1–9 (2015). https://doi.org/10.1061/(ASCE)LA.1943-4170.0000160
11. Tomar, A., Bansal, V.K.: Generation, visualization, and evaluation schedule of repetitive construction projects using GIS. Int. J. Constr. Manage., 1–16 (2019). https://doi.org/10.1080/15623599.2019.1683691
12. Martins, S.S., Evangelista, A.C.J., Hammad, A.W.A., Tam, V.W.Y., Haddad, A.: Evaluation of 4D BIM tools applicability in construction planning efficiency. Int. J. Constr. Manage. **22**(15), 2987–3000 (2022). https://doi.org/10.1080/15623599.2020.1837718
13. Kersten, G.E., Lai, L.: Negotiation support and e-negotiation systems: an overview. Group Decis. Negot. **16**, 553–586 (2007). https://doi.org/10.1007/s10726-007-9095-5
14. Dobrijevic, G., Djokovic, F.: E-negotiation: can artificial intelligence negotiate better deals? In: International Scientific Conference on Information Technology and Data Related Research, SINTEZA 2020, pp. 289–294 (2020). https://doi.org/10.15308/Sinteza-2020-289-294
15. Braun, P., et al.: e-*Negotiation* systems and software agents: methods, models, and applications. In: Intelligent Decision-making Support Systems. Decision Engineering. Springer, London (2000). https://doi.org/10.1007/1-84628-231-4_15

16. Gao, Z., Qian, Q.: The risk and benefits of applying artificial intelligence in business discussions. BCP Bus. Manag. FMEME2022 **30** (2022). https://doi.org/10.54691/bcpbm.v30i.2569
17. Lima, G.H.A., Costa, A.P.C.S.: Decision support system for maturity assessment in asset management. In: Proceedings 24th International Conference on Group Decision and Negotiation & 10th International Conference on Decision Support System Technology, Porto, Portugal (2024)
18. Martins, C.L., Neto, J.B.S.S., da Silva, L., Frej, E., de Almeida, A.T.: A GIS-based decision support system for natural gas distribution planning. In: Proceedings 24th International Conference on Group Decision and Negotiation & 10th International Conference on Decision Support System Technology, Porto, Portugal (2024)
19. Williams, C., Fang, L.: Development of a decision support system for family law in Ontario: framing a family legal case. In: Proceedings 23rd International Conference on Group Decision and Negotiation, Tokyo, Japan, 11–15 June 2023
20. Zhu, Z., Kilgour, D.M., Hipel, K.W.: Conflict analysis of offshore oil exploration in the South China sea. In: Proceedings 17th International Conference on Group Decision and Negotiation, Stuttgart, Germany (2017)
21. Covaci, F.L.: A multi-agent negotiation support system for supply chain formation. In: Proceedings 17th International Conference on Group Decision and Negotiation, Stuttgart, Germany (2017)
22. Palha, R.P., de Almeida, A.T., Morais, D.C.: A group decision and negotiation framework for hiring subcontractors in the civil construction industry. In: Proceedings 17th International Conference on Group Decision and Negotiation, Stuttgart, Germany (2017)
23. Kersten, G.E., Lo, G.: Negotiation support systems and software agents in e-business negotiations. In: Proceedings 1st International Conference on Electronic Business, Hong Kong, 19–21 December 2001
24. San Cristóbal, J.R.: The use of game theory to solve conflicts in the project management and construction industry. Int. J. Inf. Syst. Proj. Manag. **3**, 43–58 (2015)
25. Piraveenan, M.: Applications of game theory in project management: a structured review and analysis. Mathematics **7**, 858 (2019). https://doi.org/10.3390/math7090858
26. Nash, J.F., Jr.: The bargaining problem. Econom. J. Econom. Soc. **18**, 155–162 (1950)
27. Branzei, R., Dimitrov, D., Tijs, S.: Models in Cooperative Game Theory, vol. 556. Springer Science & Business Media, Berlin, Germany (2008)
28. Estévez-Fernández, A.: A game-theoretical approach to sharing penalties and rewards in projects. Eur. J. Oper. Res. **216**, 647–657 (2012)
29. San Cristóbal, J.R.: Cost allocation between activities that have caused delays in a project using game theory. Proced. Technol. **16**, 1017–1026 (2014)
30. Alnuaimi, A.S., Taha, R.A., Mohsin, M.A., Al-Harthi, A.S.: Causes, effects, benefits, and remedies of change orders on public construction projects in Oman. J. Constr. Eng. Manage. **136**(5), 615–622 (2010). https://doi.org/10.1061/(ASCE)CO.1943-7862.0000154
31. Ergin, A., Acar, E.: Change management in construction: a systematic review. In: Proceedings 7th Construction project configuration management Conference (IPCMC2022) (2022)
32. Barlish, K., Sullivan, K.: How to measure the benefits of BIM—a case study approach. Autom. Constr. **24**, 149–159 (2012). https://doi.org/10.1016/j.autcon.2012.02.008
33. Kalach, M., Abdul-Malak, M.A., Srour, I.: BIM-enabled streaming of changes and potential claims induced by fast-tracking design-build projects. J. Leg. Aff. Dispute Resolut. Eng. Constr. **13**(1), 04520042 (2021). https://doi.org/10.1061/(ASCE)LA.1943-4170.0000450
34. Ali, B., Zahoor, H., Nasir, A.R., Maqsoom, A., Khan, R.W.A., Mazher, K.M.: BIM-based claims management system: a centralized information repository for extension of time claims. Autom. Constr. **110**, 102937 (2020). https://doi.org/10.1016/j.autcon.2019.102937

35. Marzouk, M., Othman, A., Enaba, M., Zaher, M.: Using BIM to identify claims early in the construction industry: case study. J. Leg. Aff. Dispute Resolut. Eng. Constr. **10**(3), 05018001 (2018). https://doi.org/10.1061/(ASCE)LA.1943-4170.0000254
36. FIDIC conditions of contract for construction: Multilateral development bank harmonised edition, FIDIC Pink Book, Version 3 (2010)
37. Khan, M.T., Horita, M.: Building Information Model (BIM) and Geotechnical Baseline Report (GBR) for improving project management tools of underground works. In: International Conference on Construction Engineering and Project Management, pp. 532–539 (2024). https://doi.org/10.6106/ICCEPM.2024.0532
38. Acharya, N.K., Lee, Y.D., Im, H.M.: Conflicting factors in construction projects: Korean perspective. Eng. Constr. Archit. Manage. **13**(6), 543–566 (2006). https://doi.org/10.1108/09699980610712364
39. Khan, M.T., Horita, M.: Uncertainty and information asymmetry in underground works: a case study. In: Campos Ferreira, M., Wachowicz, T., Zaraté, P., Maemura, Y. (eds.) Human-Centric Decision and Negotiation Support for Societal Transitions. GDN 2024. LNBIP, vol. 509. Springer, Cham (2024). https://doi.org/10.1007/978-3-031-59373-4_2
40. Norwegian Tunneling Society: Contracts in Norwegian Tunneling. Publication No. 21 (2024)
41. Awwad, R., Barakat, B., Menassa, C.: Understanding dispute resolution in the middle east region from perspectives of different stakeholders. J. Manage. Eng. **32**(6), 05016019 (2016). https://doi.org/10.1061/(ASCE)ME.1943-5479.0000465
42. Schoop, M., Jertila, A., List, T.: Negoisst: a negotiation support system for electronic business-to-business negotiations in e-commerce. Data Knowl. Eng. **47**, 371–401 (2003). https://doi.org/10.1016/S0169-023X(03)00065-X
43. Craveiro, F., Duarte, J.P., Bartolo, H., Bartolo, P.J.: Additive manufacturing as an enabling technology for digital construction: a perspective on construction 4.0. Autom. Constr. **103**, 251–267 (2019). https://doi.org/10.1016/j.autcon.2019.03.011
44. Hardin, B., McCool, D.: BIM and Construction Management: Proven Tools, Methods, and Workflows, 2nd edn. Wiley, New York (2015)
45. Giwa, F., Omotayo, T., Tzortzopoulos, P., Malalgoda, C.: BIM-enabled claims management concept: implications for dispute avoidance and management. J. Leg. Aff. Dispute Resolut. Eng. Constr. **16**(3), 04524009 (2024). https://doi.org/10.1061/JLADAH.LADR-1112
46. Bakhary, N.A., Adnan, H., Ibrahim, A.: A study of construction claim management problems in Malaysia. Proc. Econom. Fin. **23**, 63–70 (2015)
47. CICRP: BIM project execution planning guide - Version 2.1. The Pennsylvania State University, University Park, PA, USA (2011)
48. Ibraheem, R.A.R., Mahjoob, A.M.R.: Facilitating claims settlement using building information modeling in school building projects. Innov. Infrastruct. Solut. **7**, 40 (2022). https://doi.org/10.1007/s41062-021-00646-2
49. Project Documents: Approval Documents, Bidding Documents, Contract Documents, Construction Drawings, Progress Reports and Internal Documentation (2023)

Author Index

A
Alcantud, José Carlos R. 82

B
Balle, Max 3

C
Chebotarev, Pavel 34

D
da Costa, Sergio Eduardo Gouvea 53
Danielson, Mats 68
de Souza, Pedro Henrique Gouveia 53
dos Santos Silva, Elton César 136

F
Fang, Liping 136
Frej, Eduarda Asfora 53

G
Ge, Bingfeng 124
Gu, Tianyang 124

H
Han, Sining 124
Hipel, Keith W. 111
Horita, Masahide 165

K
Kesting, Peter 18
Khan, Muhammad Tajammal 165
Kilgour, D. Marc 111
Kröcher, Felix 18

L
Lakmayer, Sebastian 68
Liu, Zihui 124

M
Maemura, Yu 150
Meyer, Marlene 3
Monteiro, Nathália Jucá 53
Morais, Danielle Costa 136

O
Ozawa, Kazumasa 150

R
Roszkowska, Ewa 94

S
Santos-García, Gustavo 82
Schoop, Mareike 3
Smolinski, Remigiusz 18

W
Wachowicz, Tomasz 94
Wang, Chi 124
Wei, Wanying 124

X
Xu, Kai 150

Z
Zhu, Ziming 111

The manufacturer's authorised representative in the EU is Springer Nature Customer Service Centre GmbH, Europaplatz 3, 69115 Heidelberg, Germany. If you have any concerns regarding our products, please contact ProductSafety@springernature.com

Printed and bound by CPI Group (UK) Ltd, Croydon, CR0 4YY

26/03/2026

02078962-0002